Robust Control of Time-delay Systems

Qing-Chang Zhong

Robust Control of Time-delay Systems

With 79 Figures

Springer

Qing-Chang Zhong, PhD
Department of Electrical Engineering and Electronics
The University of Liverpool
Brownlow Hill
Liverpool
L69 3GJ
UK

British Library Cataloguing in Publication Data
Zhong, Qing-Chang
　Robust control of time-delay systems
　1.Robust control 2.Time delay systems
　I.Title
　629.8'312
ISBN-10: 1846282640

Library of Congress Control Number: 2006921167

ISBN-10: 1-84628-264-0   e-ISBN 1-84628-265-9   Printed on acid-free paper
ISBN-13: 978-1-84628-264-5

© Springer-Verlag London Limited 2006

MATLAB® and Simulink® are the registered trademarks of The MathWorks, Inc., 3 Apple Hill Drive Natick, MA 01760-2098, U.S.A. http://www.mathworks.com

Apart from any fair dealing for the purposes of research or private study, or criticism or review, as permitted under the Copyright, Designs and Patents Act 1988, this publication may only be reproduced, stored or transmitted, in any form or by any means, with the prior permission in writing of the publishers, or in the case of reprographic reproduction in accordance with the terms of licences issued by the Copyright Licensing Agency. Enquiries concerning reproduction outside those terms should be sent to the publishers.

The use of registered names, trademarks, etc. in this publication does not imply, even in the absence of a specific statement, that such names are exempt from the relevant laws and regulations and therefore free for general use.

The publisher makes no representation, express or implied, with regard to the accuracy of the information contained in this book and cannot accept any legal responsibility or liability for any errors or omissions that may be made.

Printed in Germany

9 8 7 6 5 4 3 2 1

Springer Science+Business Media
springer.com

This book is dedicated to
SHUHONG, LILLY and LISA.

Preface

Systems with delays frequently appear in engineering. Typical examples of time-delay systems are communication networks, chemical processes, teleoperation systems, biosystems, underwater vehicles and so on. The presence of delays makes system analysis and control design much more complicated. During the last decade, we have witnessed significant development in the robust control of time-delay systems. The aim of this book is to present a systematic and comprehensive treatment of robust (H^∞) control of such systems in the frequency domain. The emphasis is on systems with a single input/output delay, although the delay-free part of the plant can be multi-input–multi-output (MIMO), when the delays in different channels are the same.

This book collects work carried out recently by the author in this field. It covers the whole range of robust (H^∞) control of time-delay systems: from controller parameterisation, controller design to controller implementation; from the Nehari problem, the one-block problem to the four-block problem; from theoretical developments to practical issues. The major tools used in this book are similarity transformations, chain-scattering approach and J-spectral factorisations. The main idea is to "*make everything as simple as possible, but not simpler (Albert Einstein).*" This book is self-contained and should be of interest to final-year undergraduates, graduates, engineers, researchers, and mathematicians who work in the area of control and time-delay systems.

The book is divided into two parts: Controller Design (Chapters 2–10) and Controller Implementation (Chapters 11–13). The classical control of time-delay systems is summarised in Chapter 2 and then some mathematical preliminaries are collected in Chapter 3. The J-spectral factorisation of regular para-Hermitian transfer functions is developed in Chapter 4 to prepare for the solution of the Nehari problem discussed in Chapter 5. An extended Nehari problem is solved in Chapter 6 to prepare for the solutions of the one-block problem and the standard H^∞ control problem discussed in Chapter 7, where the chain-scattering approach is applied to reduce the standard H^∞ control problem to a delay-free problem and a one-block problem. The latter is then further reduced to an extended Nehari problem. With the solution to the extended Nehari problem obtained in Chapter 6, the controllers for the one-block problem and the standard H^∞ problem are recovered. A transformed standard H^∞ problem is discussed in Chapter 8 to obtain a simpler but more conservative solution. The parameterisation of all stabilising controllers for time-delay systems are discussed in Chapter 9. All the controllers for the above problems have the same structure: incorporating a modified Smith predictor (MSP). A practical issue, a numerical problem with the MSP, is discussed in Chapter 10 and a unified Smith predictor is proposed to overcome this, followed by revisiting some well-studied problems. Another practical issue, the implementation of MSP, is tackled in Part II. The implementation of MSP, *i.e.*, a distributed delay, is not trivial because of the inherent hidden unstable poles. In Chapter 11, this is done by using discrete delays in the z-domain and in the s-domain. In Chapter 12, this is done by using rational transfer functions based on the δ-operator and then in Chapter 13 a faster converging rational implementation is discussed using bilinear transformations.

It is a pleasure to express my gratitude to G. Weiss, K. Gu, G. Meinsma and L. Mirkin. Special thanks go to O. Jackson (the Editor), S. Moosdorf (the Production Editor) and M. Saunders (the Copy Editor) for their professional and efficient editorial work on this book. There are no words that suffice to thank my wife Shuhong Yu for her endurance, love, support and sacrifice for my research over years. I am also grateful for financial support for my research from the Engineering and Physical Sciences Research Council (EPSRC), UK under Grant No. EP/C005953/1.

Liverpool, United Kingdom
January 2006

Qing-Chang Zhong
http://come.to/zhongqc

Contents

Notation and Abbreviations xv

List of Figures xix

List of Tables xxi

1 Introduction 1
   1.1 What Is a Delay? 1
   1.2 Examples of Time-delay Systems 2
      1.2.1 Shower 2
      1.2.2 Chemical Processes 3
      1.2.3 Communication Networks 3
      1.2.4 Underwater Vehicles 4
      1.2.5 Combustion Systems 6
      1.2.6 Exhaust Gas Recirculation (EGR) Systems 7
      1.2.7 Biosystems 8
   1.3 A Brief Review of the Control of Time-delay Systems .. 9
   1.4 Overview of This Book 10

Part I Controller Design

2 Classical Control of Time-delay Systems 17
   2.1 PID Control 17
      2.1.1 Structure of PID Controllers 17
      2.1.2 Tuning Methods for PID Controllers 18
      2.1.3 Simulation Examples 21
   2.2 Smith Predictor (SP)-based Control 22
      2.2.1 Control Difficulties Due to Delay 22
      2.2.2 Smith Predictor 24
      2.2.3 Robustness 25

		2.2.4	Disturbance Rejection	27
		2.2.5	Simulation Examples	28
	2.3	Modified Smith Predictor (MSP)-based Control		30
		2.3.1	Modified Smith Predictor	30
		2.3.2	Zero Static Error	31
		2.3.3	Simulation Examples	32
	2.4	Finite-spectrum Assignment (FSA)		39
	2.5	Connection Between MSP and FSA		39
		2.5.1	All Stabilising Controllers for Delay Systems	39
		2.5.2	Predictor–Observer Representation: MSP	40
		2.5.3	Observer–Predictor Representation: FSA	41
		2.5.4	Some Remarks	43
3	**Preliminaries**			**45**
	3.1	FIR Operators		45
	3.2	Chain-scattering Approach		46
		3.2.1	Representations of a System: IOR and CSR	46
		3.2.2	Linear Fractional Transformations: The Standard LFT and the HMT	48
		3.2.3	Some Important Properties	49
	3.3	State-space Operations on Systems		50
		3.3.1	Operations on Systems	51
		3.3.2	Similarity Transformations	57
	3.4	Algebraic Riccati Equations		58
		3.4.1	Definitions	58
		3.4.2	Stabilising Solution	58
		3.4.3	Block-diagram Representation	60
		3.4.4	Similarity Transformations and Stabilising Solutions	61
		3.4.5	Rank Defect of Stabilising Solutions	67
		3.4.6	Stabilising or Grouping?	68
	3.5	The Σ Matrix		68
		3.5.1	Definition of the Σ Matrix	69
		3.5.2	Important Properties of Σ	70
	3.6	The $L_2[0,h]$-induced Norm		72
4	***J*-spectral Factorisation of Regular Para-Hermitian Transfer Matrices**			**73**
	4.1	Introduction		73
	4.2	Properties of Projections		74
	4.3	Regular Para-Hermitian Transfer Matrices		75
	4.4	*J*-spectral Factorisation of the Full Set		77
		4.4.1	Via Similarity Transformations with Two Matrices	77
		4.4.2	Via Similarity Transformations with One Matrix	78
	4.5	*J*-spectral Factorisation of a Smaller Subset		79
	4.6	*J*-spectral Factorisation of $\Lambda = G^\sim JG$ with Stable G		82

			Contents	xi

    4.7   Numerical Examples 84

        4.7.1 $\Lambda(s) = \begin{bmatrix} 0 & \frac{s-1}{s+1} \\ \frac{s+1}{s-1} & 0 \end{bmatrix}$ 84

        4.7.2 $\Lambda(s) = \begin{bmatrix} -\frac{s^2-4}{s^2-1} & 0 \\ 0 & \frac{s^2-1}{s^2-4} \end{bmatrix}$ 84

    4.8   Summary.. 86

5   The Delay-type Nehari Problem 87
    5.1   Introduction 87
    5.2   Problem Statement (NP$_h$) 88
    5.3   Solution to the NP$_h$ 89
    5.4   Proof 90
    5.5   Special Cases 93
        5.5.1   The Stable Case.................................. 93
        5.5.2   The Conventional Nehari Problem 93
        5.5.3   The Conventional Nehari Problem with Stable A 94
    5.6   Realizations of Θ^{-1} and Θ................................. 94
    5.7 J-spectral Co-factor of Θ^{-1} 96
    5.8   A Numerical Example................................... 99
        5.8.1   The Stable Case ($a<0$)............................100
        5.8.2   The Unstable Case ($a>0$) 103
    5.9   Summary and Notes 108

6   An Extended Nehari Problem 109
    6.1   Problem Statement 109
    6.2   The Solvability Condition 110
    6.3   Solution 110
    6.4   Proof 111
        6.4.1   Rationalisation by Z_1 111
        6.4.2   Completing the J-losslessness 112
    6.5   Realization of M 113
    6.6   Summary...116

7   The Standard H^∞ Problem 117
    7.1   Introduction 117
    7.2   Problem Statements 119
        7.2.1   The Standard H^∞ Problem (SP$_h$).................119
        7.2.2   The One-block Problem (OP$_h$).....................119
    7.3   Reduction of the Standard Problem (SP$_h$) 120
        7.3.1   The Standard Delay-free H^∞ Problem (SP$_0$) 120
        7.3.2   Reducing SP$_h$ to OP$_h$ 121
        7.3.3   Reducing OP$_h$ to ENP$_h$ 122
    7.4   Solutions 124
        7.4.1   Solution to OP$_h$ 124
        7.4.2   Solution to SP$_h$ 124

xii   Contents

  7.5 Proof 125
       7.5.1 Recovering the Controller 125
       7.5.2 Realization of V^{-1} 126
  7.6 Summary and Notes 128

8 A Transformed Standard H^∞ Problem 129
  8.1 Introduction 129
  8.2 The Transformation 130
  8.3 Solution 132
  8.4 A Numerical Example 135
  8.5 Summary 138

9 2DOF Controller Parameterisation 139
  9.1 Parameterisation of the Controller 139
  9.2 Two-degree-of-freedom Realization of the Controller 142
       9.2.1 Control Structure 142
       9.2.2 Set-point Response 143
       9.2.3 Disturbance Response 144
       9.2.4 Robustness Analysis 145
       9.2.5 Ideal Disturbance Response 145
       9.2.6 Realization of $P_1 - N\tilde{Y}P$ 147
  9.3 Application to Integral Processes with Dead Time 148
  9.4 Summary 151

10 Unified Smith Predictor 153
  10.1 Introduction 153
  10.2 Predictor-based Control Structure 154
  10.3 Problem Identification and the Solution 156
        10.3.1 A Numerical Problem with the MSP 156
        10.3.2 The Unified Smith Predictor (USP) 156
  10.4 Control Systems with a USP: Equivalent Diagrams 160
  10.5 Applications 164
        10.5.1 Parameterisation of All Stabilising Controllers .. 164
        10.5.2 The H^2 Problem 165
        10.5.3 A Transformed H^∞ Problem 169
  10.6 Summary 170

Part II Controller Implementation

---

11 Discrete-delay Implementation of Distributed Delay in Control Laws 173
  11.1 Introduction 173
  11.2 A Bad Approximation of Distributed Delay in the Literature .. 175
  11.3 Approximation of Distributed Delay 176

11.3.1 Integration $\int_0^{\frac{h}{N}} y(t-\tau)d\tau$ 176
11.3.2 Approximation in the s-domain via the
Laplace Transform................................ 177
11.3.3 Direct Approximation in the s-domain 180
11.3.4 Equivalents for the Backward Rectangular Rule 182
11.4 Implementation of Distributed Delay Z 183
11.4.1 Implementation of Z in the z-domain 183
11.4.2 Implementation of ZOH in the s-domain 184
11.4.3 Implementation of Z in the s-domain 185
11.5 Stability Issues Related to the Implementation 187
11.6 Numerical Examples 188
11.6.1 Approximations and Implementations of
Distributed Delay 188
11.6.2 System Responses using Different Implementations 191
11.6.3 Numerical Integration Using the Improved
Rectangular Rules 193
11.7 Summary... 194

12 Rational Implementation Inspired by the δ-operator 195
12.1 Introduction 195
12.2 The δ-operator 196
12.3 An Initial Approximation................................ 196
12.4 Implementation with Zero Static Error 197
12.5 Convergence of the Implementation 201
12.6 Structure of the Implementation.......................... 203
12.7 Numerical Examples 204
12.8 Summary... 205

**13 Rational Implementation Based on the Bilinear
Transformation** 207
13.1 Preliminary: Bilinear Transformation 207
13.2 Implementation of Distributed Delay...................... 208
13.3 Convergence of the Implementation 211
13.4 Numerical Examples 214
13.5 Summary... 216

References... 217

Index.. 229

Notation and Abbreviations

$\mathbb{Z}$, $\mathbb{R}$ and $\mathbb{C}$	fields of integral, real and complex numbers
$j\mathbb{R}$	imaginary axis
Re s and Im s	real and imaginary parts of $s \in \mathbb{C}$
$\in$	belong to
$\cap$	intersection
$\subset$	subset
I_n	$n \times n$ identity matrix (n is often omitted when not confusing)
$J_{p,q}$, J_γ and J	shorthand for $\begin{bmatrix} I_p & 0 \\ 0 & -I_q \end{bmatrix}$, $\begin{bmatrix} I & 0 \\ 0 & -\gamma^2 I \end{bmatrix}$ and $\begin{bmatrix} I & 0 \\ 0 & -I \end{bmatrix}$
A^T and A^*	transpose and complex conjugate transpose of A
A^{-1} and A^{-*}	inverse of A and shorthand for $(A^{-1})^*$
$\det(A)$ and $\rho(A)$	determinant and spectral radius of A
Im A	image of A
$\left[\begin{array}{c\|c} A & B \\ \hline C & D \end{array}\right]$	explicitly partitioned matrix of $\begin{bmatrix} A & B \\ C & D \end{bmatrix}$
$G(s) = \left[\begin{array}{c\|c} A & B \\ \hline C & D \end{array}\right]$	shorthand for $G(s) = D + C(sI - A)^{-1}B$
$G^\sim(s)$	shorthand for $G^T(-s) = [G(-s^*)]^* = \left[\begin{array}{c\|c} -A^* & -C^* \\ \hline B^* & D^* \end{array}\right]$
$\mathcal{F}_l(M, Q)$ and $\mathcal{F}_u(M, Q)$	lower and upper linear fractional transformations (LFT)
$\mathcal{H}_r(M, Q)$ and $\mathcal{H}_l(M, Q)$	right and left homographic transformations (HMT)
$\mathcal{C}_r(M)$ and $\mathcal{C}_l(M)$	right and left chain-scattering transformations (CST)
$\pi_h(G)$	completion operator
$\tau_h(G)$	truncation operator

ARE	algebraic Riccati equation
w.r.t.	with respect to
iff	if and only if
IOR	input–output representation
CSR	chain-scattering representation
LFT	linear fractional transformation
HMT	homographic transformation
SP	(classical) Smith predictor
MSP	modified Smith predictor
USP	unified Smith predictor
FSA	finite-spectrum assignment
FIR	finite impulse response
SP_h	standard H^∞ problem with a single delay h
SP_0	the conventional standard H^∞ problem without a delay ($h=0$)
OP_h	one-block problem with a delay h
ENP_h	extended Nehari problem with a delay h
NP_h	delay type Nehari problem
SISO	single-input single-output
MIMO	multiple-input multiple-output
ZOH	zero-order hold

List of Figures

1.1	Sketch of a shower system	3
1.2	Communication networks: A single connection	4
1.3	The MIT underwater vehicle: Odyssey II, Xanthos	5
1.4	Physical underwater vehicle system	6
1.5	Dynamics in a combustion system	7
2.1	Unity negative feedback control system	20
2.2	PID control: Example 1	22
2.3	PID control: Example 2	23
2.4	SP-based control system	24
2.5	SP-based control system: Internal model control	25
2.6	SP-based control system: Nominal case and $d = 0$	25
2.7	SP Example 1: Nominal responses	28
2.8	SP Example 1: Robustness	29
2.9	SP Example 2: Unstable plant	30
2.10	MSP Example 1: Stability region of (K_p, T_i)	33
2.11	Root-locus design of PI controller for MSP-based stable system	33
2.12	Performance of the MSP-based stable system: $K_p = 7$, $T_i = 2$	34
2.13	Performance of the MSP-based stable system: $K_p = 7$, $T_i = 0.2$	35
2.14	MSP Example 2: Stability region of (K_p, T_i)	36
2.15	Root-locus design for MSP-based unstable system	37
2.16	Performance of the MSP-based unstable system	38
2.17	Stabilising controllers for processes with dead time	40
2.18	Predictor–Observer Representation: MSP scheme	42
2.19	Observer–Predictor Representation: FSA scheme	42
3.1	Completion operator π_h and truncation operator τ_h	46
3.2	Input–output representation of a system	47
3.3	Chain-scattering representations of the system M in Figure 3.2	48
3.4	State feedback $u = Fx + v$	53
3.5	Output injection	53

3.6	Star product of interconnected systems	54
3.7	Similarity transformation on a system G with T	57
3.8	Block-diagram representation of algebraic Riccati equation	61
3.9	Solution generator when $T = \begin{bmatrix} I & 0 \\ L & I \end{bmatrix}$	63
3.10	Solution generator when $T = \begin{bmatrix} L & 0 \\ 0 & L^{-T} \end{bmatrix}$	64
3.11	Solution generator when $T = \begin{bmatrix} I & L \\ 0 & I \end{bmatrix}$	65
3.12	Solution generator when $T = \begin{bmatrix} 0 & I \\ I & 0 \end{bmatrix}$	66
3.13	Solution generator when $T = \begin{bmatrix} -I & 0 \\ 0 & I \end{bmatrix}$	67
5.1	Representation of the NP_h as a block diagram	90
5.2	Surface of $\hat{\Sigma}_{22}$ with respect to ah and $a\gamma$ ($a < 0$)	102
5.3	Contour $\hat{\Sigma}_{22} = 0$ on the ah-$a\gamma$ plane ($a < 0$)	102
5.4	Locus of the hidden poles of Z	103
5.5	Nyquist plots of Π_{22} ($a < 0$ and $ah = -1$)	104
5.6	Surface of $\hat{\Sigma}_{22}$ with respect to ah and $a\gamma$ ($a > 0$)	106
5.7	Contour $\hat{\Sigma}_{22} = 0$ on the ah-$a\gamma$ plane ($a > 0$)	106
5.8	Nyquist plots of Π_{22} ($a > 0$ and $ah = 1$)	107
6.1	Solving the ENP_h	111
7.1	General setup of control systems with a single I/O delay	119
7.2	Reduction of SP_h to SP_0 and a one-block delay problem	122
7.3	Reducing the one-block problem to ENP_h	123
7.4	Recovering the controller K	127
8.1	General control setup for dead-time systems	131
8.2	An equivalent structure	131
8.3	Graphic interpretation of the transformation	132
8.4	Setup for mixed sensitivity minimisation	135
8.5	Comparison of $\|T_{zw}(j\omega)\|$	136
8.6	Singular value plots of $Z_1(s)$	137
9.1	Parameterised control structure with 2DOF	143
9.2	The dual locus to judge controller stability	151
10.1	Dead-time plant with a predictor-based controller	155
10.2	Unified Smith predictor $Z = Z_s + Z_u$	158

10.3 Implementation of the USP (10.9) for multiple delays, using two resetting LTI systems to implement Z_u. The logical controller which controls the switches and generates the resetting signals R_a and R_b (in an open-loop manner) is not shown. 160
10.4 Control system comprising a dead-time plant P_h (with a rational part P) and a stabilising controller K 161
10.5 Control system from Figure 10.4, in which the controller K has been decomposed into a USP denoted Z and a stabilising compensator C, so that $K = C(I - ZC)^{-1}$. 162
10.6 Equivalent representation of the control system in Figure 10.5, using the decomposition of P^{aug} given in Proposition 10.1. 162

11.1 Illustration of Lemma 11.1................................... 176
11.2 Magnitude coefficients (for a scalar A) 181
11.3 Implementations of Z in the z-domain 184
11.4 Impulse responses of different implementations of ZOH 186
11.5 Errors of different approximations ($N = 20$) 189
11.6 E_{b0} and E_{f0} for different N (zero error at frequencies 0 and $+\infty$) . 190
11.7 Implementation error of $Z_{f\epsilon}$ for different ϵ ($N = 1$) 191
11.8 Unit step response ($N = 8$) 192
11.9 Unit-step response: Z implemented as $Z_{f\epsilon}$ ($N = 1$) 193

12.1 Surface of $f(\sigma, \omega)$: $f(\sigma, \omega) = 0$ on the white lines 200
12.2 Contour of $f(\sigma, \omega)$ at level 0 200
12.3 Rational implementation: $Z_r = \Sigma_{k=1}^{N} \Pi^k \cdot \Phi^{-1} B$ 203
12.4 Implementation error of Z_r for different N 204
12.5 Unit-step response... 205

13.1 Rational implementation of distributed delay: $Z_r = \Sigma_{k=0}^{N-1} \Pi^k \Xi B$. 209
13.2 Circle into which all eigenvalues of $A\frac{h}{N}$ fall when $N > \tilde{N}$ 212
13.3 Area mapped from the right-half circle in Figure 13.2 via $\phi = c\frac{1+e^{-c}}{1-e^{-c}}$ 212
13.4 Implementation error of Z_r for different N 215
13.5 Comparison of different implementations ($N = 5$) 215
13.6 System responses when $r(t) = 1(t)$ 216

List of Tables

2.1 Ziegler–Nichols tuning formulas 19
2.2 PI control parameters for Example 1 21
2.3 PI control parameters for Example 2 22

3.1 CSR notation used in [58] and in this book 47
3.2 Similarities between CSR and IOR 50
3.3 Basic similarity transformation operations on a system 57

8.1 Performance comparison 136

10.1 USP needed for different types of plants 158

1
Introduction

Systems with delays abound in the world. One reason is that nature is full of transparent delays. Another reason is that time-delay systems are often used to model a large class of engineering systems, where propagation and transmission of information or material are involved. The presence of delays (especially, long delays) makes system analysis and control design much more complex. In this chapter, some examples of time-delay systems are discussed and then a brief review on the control of time-delay systems, followed by an overview of this book, is given.

1.1 What Is a Delay?

Time delay is the property of a physical system by which the response to an applied force (action) is delayed in its effect [124]. Whenever material, information or energy is physically transmitted from one place to another, there is a delay associated with the transmission. The value of the delay is determined by the distance and the transmission speed. Some delays are short, some are very long. The presence of long delays makes system analysis and control design much more complex. What is worse is that some delays are too long to perceive and the system is misperceived as one without delays.

Time delays abound in the world. They appear in various systems such as biological, ecological, economic, social, engineering systems etc. For example, over-exposure to radiation increases the risk of cancer, but the onset of cancer typically follows exposure to radiation by many years. In economics, the central bank in a country often attempts to influence the economy by adjusting interest rates; the effect of a change in interest rates takes months to be translated into an impact on the economy. In politics, politicians need some time to make decisions and they will have to wait for some time before they find out if the decisions are correct or not. When reversing a car around a corner, the driver has to wait for the steering to take effect. In engineering, on which this book focuses, there are a lot of systems with delays; see the next section for examples.

A general tendency in responding to some errors in a system is to react immediately to the errors and to react more if the errors are not lessened or eliminated in time as expected. However, for a system with time delays, only after the inherent delays will the errors start to change. Hence, it is very important to properly understand the existence of delays and not to over-react. Otherwise, the system is very likely to overshoot or even become unstable. When dealing with time-delay systems, "patience is a virtue."

For a given delay element with a delay $h \geq 0$, the output $y(t)$ corresponding to the input $u(t)$ is
$$y(t) = u(t-h).$$
Hence, the transfer function of a delay element is given by e^{-sh}.

1.2 Examples of Time-delay Systems

Some typical examples of time-delay systems in engineering are discussed here.

1.2.1 Shower

A simple example of a time-delay system from everyday life is the shower,[1] as depicted in Figure 1.1. Most people have experienced the difficulty in adjusting the water temperature: it gets too cold or too warm. The actual temperature often overshoots the desired and, sometimes, it takes a while to get the temperature right. This is because it takes time for the increased (or decreased) hot/cold water to flow from the tap to the shower head (or the human body). This time is a delay, which depends on the water pressure and the length of the pipe. The change of the faucet position is almost immediate, however, the change of the water temperature has to wait until the delay has elapsed. If the faucet position is constantly adjusted according to the currently perceived temperature, then it is very likely that the temperature will fluctuate.

Assume that the water is an incompressible fluid and stationary flow. According to the Poiseuille law, the flow rate of water is
$$F = \frac{\pi R^4}{8\mu l} \Delta p,$$
where $\mu = 0.01$ is the kinematic viscosity of water, R is the radius of the pipe, l is the length of the pipe and Δp is the pressure difference between the two ends of the pipe. The time delay h can then be found as
$$h = \frac{\pi R^2 l}{F} = \frac{8\mu}{\Delta p}\left(\frac{l}{R}\right)^2.$$

---

[1] A shower model written in SIMULINK® is available at
http://www.aut.bme.hu/education/contheor/contheor/theory/model/shower/.

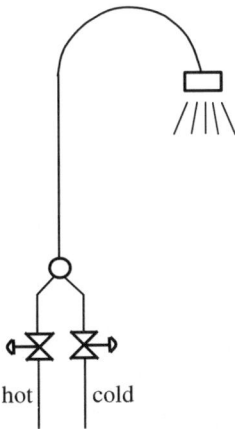

Figure 1.1. Sketch of a shower system

1.2.2 Chemical Processes

It is well known that many chemical processes contain delays. This has been well documented in the literature and there is no need to give any detailed example here. The following first-order plus dead time (FOPDT) is widely used to model chemical processes:

$$G(s) = \frac{K}{Ts+1} e^{-sh},$$

where K is the static gain of the plant, $T > 0$ is the time constant and h is the transparent delay or dead time.

1.2.3 Communication Networks

In recent years, communication networks [143] have been among the fastest-growing areas in engineering and there has been increasing interest in controlling systems over communication networks. Thanks to high-speed networks, control-over-Internet is now available [69, 122]. These systems are frequently modelled from the control point of view as time-delay systems because of the inherent propagation delays [52, 71]. These delays are crucial to the system stability and the quality-of-service (QoS).

A single connection between a source controlled by an access regulator and a distant destination node served with a constant transmission capacity μ is given as an example here. This can be described by the fluid model [52, 71] shown in Figure 1.2(a). At the source node, the access regulator controls the input rate $u(t)$, according to the congestion status of the destination node.

The congestion status $y(t)$ is defined as the difference between the current buffer contents $x(t)$ and the target value $\bar{X}$, i.e.,

$$y(t) = x(t) - \bar{X}.$$

Due to the propagation delay from the destination node to the source node, called the backward delay h_b, this status arrives at the source node (the access regulator) only after this delay period. There is also a forward delay h_f for the package to arrive at the destination node from the source node. The arrived packages are stored/accumulated in the buffer and then sent with a constant transmission capacity μ. The control objective is to adapt $u(t)$ to μ dynamically while maintaining the buffer $x(t)$ at an acceptable level. The block diagram is shown in Figure 1.2(b).

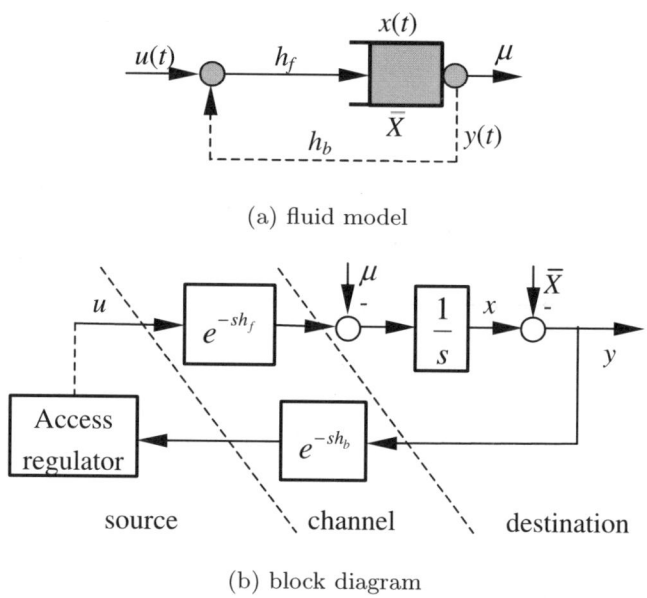

(a) fluid model

(b) block diagram

Figure 1.2. Communication networks: A single connection

The communication networks in reality, which are built from single connections, are much more complicated. The delays are often time-varying and stochastic. The information transmitted via communication networks is quantised and there exist package losses as well.

1.2.4 Underwater Vehicles

Recently, there have been more and more applications of underwater vehicles. They can be used for exploring ocean bottoms, installation/inspection/repair tasks and, of course, military missions. They have advantages over human divers in that they can descend to greater depths, can stay there for greater lengths of time and require less support equipment. Thus they can reach places divers cannot, and they can be less expensive to operate [64].

There are different types of underwater vehicles. One is the *Remotely Operated Vehicle* (ROV). An ROV is connected to a surface support ship via an umbilical cable, which provides power supply and a communication link, and hence the range of operation is somewhat limited. Another one is the *autonomous underwater vehicle* (AUV), which carries an on-board power unit and is equipped with advanced control capabilities to undertake tasks with the minimum of human intervention. The communication is carried out through an acoustic link. The MIT underwater vehicle Odyssey II Xanthos, shown in Figure 1.3, is taken as an example. Odyssey II Xanthos is a video survey AUV, equipped with various sensors including scanning and homing sonars, depth sensor, temperature salinity and related sensors, video, inertial sensors, acoustic modem and acoustic navigation tracking pingers. The rated operating depth is 3,000 m. More details can be found at http://auvlab.mit.edu.

Figure 1.3. The MIT underwater vehicle: Odyssey II, Xanthos

(Courtesy of C. Chryssostomidis and R. Damus, MIT Sea Grant AUV Lab, http://auvlab.mit.edu)

The control problems involved in these vehicles include navigation, task planning and the low-level autopilot. Due to the long cable or distance, there exists a long delay in the system. For AUVs, the delay is caused by the finite sound speed in water, nominally, 1, 500 m/s.

A physical system [140] is shown in Figure 1.4, where a surface ship is shown positioning an underwater vehicle through a long cable. The vehicle may be searching the ocean floor or mapping the topography of the bottom, or it may be the platform for a smaller vehicle equipped with thrusters. An

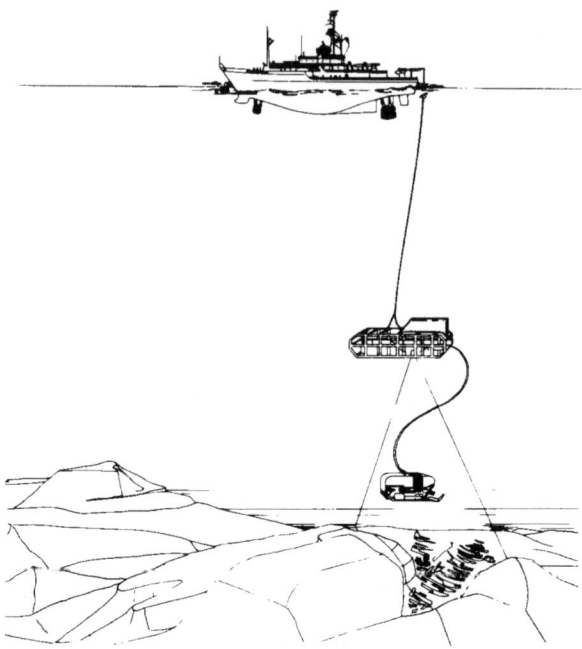

Figure 1.4. Physical underwater vehicle system

(Reprinted, with permission, from [140]. ©IEEE)

approximate model, which was validated for a cable of $2,500$ m and a vehicle weighing $17,000$ N in air, was given in [140] as

$$G(s) = \frac{ce^{-sh}}{as^2 + bs + c},$$

with $a = 1$, $b = 1.1 \times 10^{-4}$, $c = 2.58 \times 10^{-2}$ and $h = 40$ s. More details about this system can be found in [140] and the references therein.

1.2.5 Combustion Systems

Continuous combustion systems are widely used in power generation, heating and propulsion. Examples include domestic and industrial burners, steam and gas turbines, waste incinerators, and jet and ramjet engines. These systems are intricate and include a wide variety of dynamic behaviour. Pressure oscillations are considered the most significant in terms of the impact on system performance; much effort has been devoted to this [3, 24, 103].

There are two major dynamics in a combustion system: flame dynamics and acoustic wave dynamics. They are coupled to form a loop as shown in

Figure 1.5. Due to wave propagation, there is a delay in the wave dynamics. This often causes combustion instability [3, 24, 103]. Delays also appear in the measurement and the actuator units of the system. Detailed modelling of combustion systems can be found in [3, 24] and are omitted here.

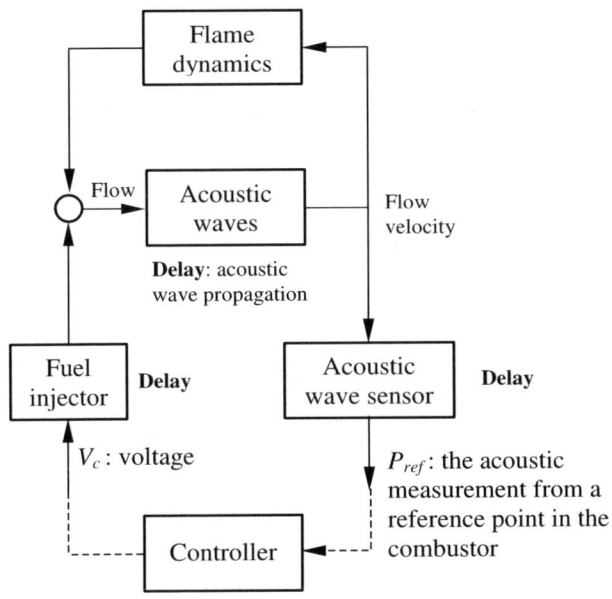

Figure 1.5. Dynamics in a combustion system

1.2.6 Exhaust Gas Recirculation (EGR) Systems

Oxides of nitrogen (NO_x) are a group of highly reactive gases that contain varying amounts of NO and NO_2. These are key elements of greenhouse gases and harmful to human health and the environment. Motor vehicles are one of the key sources of NO_x. Exhaust gas recirculation (EGR) systems were introduced in the early 1970s to reduce NO_x emissions.

The EGR valve recirculates exhaust into the intake stream. Exhaust gases have already combusted, so they do not burn again when they are recirculated. These gases displace some of the normal intake charge. This chemically slows and cools the combustion process by several hundred degrees, thus reducing NO_x formation.[2]

It is a challenge to precisely control the flow of recirculated exhaust so that the system provides good performance and economy. Too much flow will retard

---

[2] http://www.asashop.org/autoinc/nov97/gas.htm.

engine performance and cause a hesitation on acceleration, and too little flow will increase NO_x and cause engine ping. Some EGR systems simply operate in open loop due to nonlinearity, engine vibrations, pressure fluctuations and high-order unmodelled dynamics [109]. The measurement, *e.g.*, of oxygen in the exhaust [1] for feedback, often introduces delay, which complicates the control design. On the other hand, it is very difficult to derive a mathematical model for the system. The model is often obtained via system identification. The following model was given in [109]:

$$G_i(z) = k(z-p)z^{-d} \qquad (i=1,2,3,4)$$

at four operating points dependent on load, speed and the desired EGR rate for an EGR system. With a sampling period of 10 ms, the vectors of the parameters are:

$$\text{gain } k = (88, 110, 180, 220),$$
$$\text{delay } d = (5, 15, 13, 6),$$
$$\text{pole } p = (0.9, 0.9, 0.9, 0.9).$$

The model changes dramatically at different operating points. The delay varies from 50 ms to 150 ms and the gain varies from 88 to 220.

1.2.7 Biosystems

This subsection is based on [10, 16], in which more examples and references can be found.

Time delay has been introduced to model biosystems to produce better consistency with nature and predictive results. The ultimate objective is to further understand the systems and then to control them. The biosystems studied in this field cover population dynamics, epidemiology, physiology, immunology, neural networks and cell kinetics.

The following delay model for population dynamics was introduced in [48], modifying the classical Verhulst model to account for hatching and maturation periods:

$$y'(t) = ry(t)\left(1 - \frac{y(t-\tau)}{K}\right).$$

Here, the nonnegative parameters r and K are, respectively, the intrinsic growth rate and the environmental carrying capacity. This simple model can explain the observed oscillatory behaviour in a single species population, without any predatory interaction of other species.

Another simple example is the model of growth in cell populations, which is given by

$$y'(t) = \alpha y(t) + \beta y(t-\tau).$$

The equilibrium solution $y(t) = 0$ becomes unstable when the value of the delay exceeds the following bound:

$$\bar{\tau} = \frac{\cos^{-1}(-\alpha/\beta)}{\sqrt{\beta^2 - \alpha^2}}.$$

More examples of biosystems can be found in [10, 16, 67] and references therein.

1.3 A Brief Review of the Control of Time-delay Systems

The first effective control scheme for time-delay systems is the celebrated Smith predictor [126, 127]. By introducing a minor feedback loop consisting of a predictor, the controller design problem for a time-delay system is then converted to a design problem for a delay-free system. This has considerably simplified the controller design. However, the classical Smith predictor cannot be applied to unstable systems. This motivated the modified Smith predictor [113, 147] and finite-spectrum assignment [70].

H^∞ control is a key approach to deal with robustness [32, 40, 181]. The standard H^∞ control problem for delay-free systems was solved in the late 1980s [25, 32, 39]. Since then, the robust control of time-delay systems has attracted many researchers. The approaches involved are mainly of three kinds [80]: operator-theoretic methods; state-space methods; and J-spectral factorisation methods. In the context of the operator-theoretic methods, which are often based on commutant lifting methods or skew Toeplitz theory, a time-delay system is treated as a general infinite-dimensional system. Very elegant results have been obtained in [26, 30, 31, 115, 138, 139, 182]. However, since the general framework of infinite-dimensional systems is used, the solutions are very complicated and difficult for engineers to understand. In the context of state-space methods, as reported in [59, 100, 131, 132, 133, 134], the dynamic-game theory [13] plays an important role. In the context of J-spectral factorisation methods, as reported in [75, 77, 78, 80, 88, 160, 162], the Meinsma–Zwart idea, *i.e.*, to convert the J-spectral factorisation of an infinite-dimensional transfer matrix $G^\sim JG$ to that of a finite-dimensional matrix $\Theta = \Pi^\sim G^\sim JG\Pi$, where G is the plant and Π is an infinite-dimensional unimodular matrix, plays the key role. Another contribution to this area is [85, 87], where the standard H^∞ control problem for systems with a delay is solved by regarding the delay as a causality constraint on the controller but not as part of the plant so that the controller can be extracted from the controller for the delay-free problem. Recently, Meinsma and Mirkin [75, 76, 77] have made important progress on H^∞ control of systems with multiple I/O delays. For H^∞ control of a class of more general infinite-dimensional systems, see [54, 55, 57].

The above is only a brief review of the control of time-delay systems; more detailed reviews on the stability and control of time-delay systems can be found in [42, 45, 92, 102, 121, 150]. Literature reviews for specific problems studied in this book are contained in the Introduction to the relevant chapter.

1.4 Overview of This Book

The rest of this book consists of two parts and 12 chapters.

Part I Controller Design

Chapter 2   Classical Control of Time-delay Systems

Classical control approaches for time-delay systems are summarised in this chapter. These include proportional–integral–derivative (PID) control, classical Smith predictor, modified Smith predictor and finite-spectrum assignment.

Chapter 3   Preliminaries

Some preliminaries are collected in this chapter for later use. These include two important FIR operators which map a rational transfer matrix into FIR blocks, the state-space operations of systems, the chain-scattering approach, an important matrix called the Σ matrix, and the $L_2[0, h]$-induced norm.

Chapter 4   J-spectral Factorisation of Regular Para-Hermitian Transfer Matrices

This chapter characterises a class of regular para-Hermitian transfer matrices and then studies the J-spectral factorisation of this class using similarity transformations. A transfer matrix Λ in this class admits a J-spectral factorisation if and only if there exists a common nonsingular matrix to similarly transform the A-matrices of Λ and Λ^{-1}, respectively, into 2×2 lower (upper, respectively) triangular block matrices with the $(1,1)$-block including all the stable modes of Λ (Λ^{-1}, respectively). For a transfer matrix in a smaller subset, this nonsingular matrix is formulated in terms of the stabilising solutions of two algebraic Riccati equations. The J-spectral factor is formulated in terms of the original realization of the transfer matrix. The approach developed here is used in the next chapter to solve the delay-type Nehari problem. This chapter is written based on [169, 170].

Chapter 5   The Delay-type Nehari Problem

This chapter generalises the frequency-domain results for the delay-type Nehari problem in the stable case to the unstable case. It also extends the solution to the conventional (delay-free) Nehari problem to the delay-type Nehari problem. The solvability condition of the delay-type Nehari problem is formulated in terms of the nonsingularity of a delay-dependent matrix. The optimal value γ_{opt} is the maximal $\gamma \in [0, \infty)$ such that this matrix becomes singular when γ decreases from ∞. All suboptimal compensators are parameterised in a transparent structure incorporating a modified Smith predictor. This chapter is written based on [158, 160, 168].

Chapter 6 An Extended Nehari Problem

In this chapter, a different type of Nehari problem with a delay is considered. Here, instead of the requirement for stability of the compensator K, stability of the closed-loop transfer matrix is required. Hence, the norm involved in this chapter is the H^∞-norm rather than the L^∞-norm. As will be seen in the next chapter, the solution to this problem is vital for solving the standard H^∞ problem with a delay. While the solvability condition of this problem is well known, the parameterisation of all the suboptimal compensators is not trivial. This chapter is written based on [162].

Chapter 7 The Standard H^∞ Problem

In this chapter, the standard H^∞ control problem for processes with a single delay is considered. A frequency-domain approach is proposed to split the problem to a standard delay-free H^∞ problem and a one-block problem. The one-block problem is then further reduced to an extended Nehari problem. Hence, for a given bound on the H^∞-norm of the closed-loop transfer function, there exist proper stabilising controllers that achieve this bound if and only if the corresponding delay-free H^∞ problem and the extended Nehari problem with a delay (or the one-block problem) are all solvable. Applying the results obtained in the previous chapter, the solvability conditions of the standard H^∞ control problem with a delay are then formulated in terms of the existence of solutions to two delay-independent algebraic Riccati equations and the nonsingularity property of a delay-dependent matrix. All suboptimal controllers solving the problems are, respectively, parameterised as a structure incorporating a modified Smith predictor. This chapter is written based on [162].

Chapter 8 A Transformed Standard H^∞ Problem

In this chapter, a transformation is presented to solve the standard H^∞ problem of dead-time systems similarly as in the finite-dimensional situations. With some trade-off of performance, the following advantages are obtained: (i) the controller has a quite simple and transparent structure; (ii) there are no additional hidden modes in the Smith predictor. As a result, the practical significance of the approach is obvious. This chapter is written based on [157, 161].

Chapter 9 2DOF Controller Parameterisation

In this chapter, the co-prime factorisation of all stabilising controllers is presented and then the controller is realized in a two-degree-of-freedom structure. One degree-of-freedom $F(s)$ is chosen to meet the desired set-point response and the other degree-of-freedom, the free parameter $Q(s)$, is chosen to meet the desired disturbance response and to compromise the disturbance performance with robustness. Furthermore, the controller is re-configured in the chain-scattering representation. With this structure, which is symmetrical for the process and the disturbance degree-of-freedom $Q(s)$, one can see clearly the two degrees-of-freedom and the differences between the controllers for processes with and without dead time. It is also shown that the subideal disturbance response can be obtained with suitable choice of $Q(s)$. As a special case, the method is applied to integral processes with dead time. Some of these results can be found in [93, 94].

Chapter 10   Unified Smith Predictor

As can be seen from the previous chapters, modified Smith predictors (MSP) have played a very important role in the control of time-delay systems. In this chapter, a numerical problem associated with the MSP is identified and an alternative predictor, named the *unified Smith predictor* (USP), is proposed to overcome this problem. The proposed USP combines the classical Smith predictor with the modified one, after spectral decomposition of the plant. An equivalent representation of the original delay system, together with the USP, is derived. Based on this representation, all the stabilising controllers are parameterised and the standard H^2 problem is solved. This chapter is written based on [165, 179].

Part II Controller Implementation

Chapter 11   Discrete-delay Implementation of Distributed Delay in Control Laws

As shown in previous chapters, the suboptimal controllers for the Nehari problem, the extended Nehari problem, the one-block problem and the standard problem have the same structure. They all incorporate a distributed-delay block, which is in the form of a modified Smith predictor (MSP). The implementation of distributed delay is not trivial because of the inherent hidden unstable poles. In this chapter, some elementary mathematical tools are used to approximate the distributed delay and, furthermore, to implement it in the z-domain and in the s-domain. The H^∞-norm of the approximation error converges to 0 when the approximation step N approaches $+\infty$. Hence, the instability problem due to the approximation error does not exist provided that the number N of the approximation steps is large enough. Moreover, the static gain is guaranteed in the implementation so that no extra effort is needed to retain the steady-state performance. It is recommended not to use the backward rectangular rule to approximate the distributed delay for implementation. As by-products, two new formulae for the forward and backward rectangular rules are proposed. These formulae are more accurate than the conventional ones when the integrand has an exponential term. This chapter is written based on [166, 167].

Chapter 12 Rational Implementation Inspired from the δ-operator

In this chapter, a rational implementation for distributed delay is proposed. The main benefit of doing so is the easy implementation of rational transfer functions. The proposed approach was inspired by the δ-operator. The resulting rational implementation has an elegant structure of chained low-pass filters. The stability of each node can be guaranteed by the choice of the total number N of nodes. The stability of the closed-loop system can be guaranteed because the H^∞-norm of the implementation error approaches 0 when N goes to ∞. Moreover, the steady-state performance of the system is retained without the need to change the control structure. This chapter is written based on [171].

Chapter 13 Rational Implementation Based on the Bilinear Transformation

Based on an extension of the bilinear transformation, a rational implementation for distributed delay in linear control laws is proposed. This implementation converges much faster than the rational implementation inspired from the δ-operator. The implementation has an elegant structure of chained bi-proper nodes cascaded with a strictly proper node. The stability of each node is determined by the choice of the total number N of the nodes. The H^∞-norm of the implementation error approaches 0 when N goes to ∞ and hence the stability of the closed-loop system can be guaranteed. In addition, the steady-state performance of the system is retained. Simulation examples are given to verify the results and to show comparative study with other implementations. This chapter is written based on [172, 173].

In this book, much effort has been made to solve complicated problems using simple ideas and elementary mathematical tools.

Every important idea is simple.

War and Peace,   Count L.N. Tolstoy.

Part I

Controller Design

2
Classical Control of Time-delay Systems

The classical control approaches for time-delay systems are summarised in this chapter. These include proportional–integral–derivative (PID) control, classical Smith predictor, modified Smith predictor and finite-spectrum assignment.

2.1 PID Control

2.1.1 Structure of PID Controllers

A PID controller consists of three terms/modes/actions: proportional, integral and derivative. Different combinations of these terms result in different controllers, such as PI controllers and PD controllers. The standard form of a PID controller is given in the s-domain as

$$C(s) = \text{P} + \text{I} + \text{D} = K_p + \frac{K_i}{s} + K_d s,$$

where K_p, K_i and K_d are called the proportional gain, the integral gain and the derivative gain respectively. In the time domain, the output of the PID controller u can be described as follows:

$$u(t) = K_p e(t) + K_i \int e(t)dt + K_d \frac{\text{d}e(t)}{\text{d}t},$$

where $e(t)$ is the input to the controller. The following form of the PID controller is also frequently used:

$$C(s) = K_p(1 + \frac{1}{T_i s} + T_d s), \qquad (2.1)$$

where T_i is called the integral time constant or reset time and T_d is called the derivative time constant or rate time.

The PID controller is very simple and can easily be implemented using pneumatic devices, hydraulic devices, mechanical devices, electronic devices, and of course software. It takes the past (I), present (P) and future (D) information of the control error into account so that in many cases it is able to provide satisfactory control performance. Moreover, PID controllers are very robust to plant uncertainties. As a matter of fact, more than 90% industrial processes are controlled by PID controllers, mostly PI controllers [7]. PID controllers are also widely used with time-delay systems. There are many references about various issues of PID controllers, such as noise filtering and high frequency roll-off, set-point weighting and two-degree-of-freedom, windup, tuning and computer implementation; see, for example, [2, 5, 6, 7, 18, 46, 136, 155].

2.1.2 Tuning Methods for PID Controllers

A PID controller has three parameters and can be tuned by many methods, such as trial-and-error tuning, empirical tuning like the well-known Ziegler–Nichols method, analytical tuning, optimised tuning and auto-tuning with identification of the plant model [2, 5, 6, 7, 18, 37, 46, 136, 155]. Here, only some general guidelines for trial-and-error tuning, the Ziegler–Nichols method, and an analytical tuning method based on gain and phase margins will be discussed.

Trial-and-error Tuning

The impact of the three parameters on control performance is very complicated and interdependent. The following general guidelines are often used to manually tune a PID controller for an open-loop stable system:

- The proportional term provides an immediate action in the control signal corresponding to the control error. Hence, the larger the proportional gain, the faster the response. Moreover, the larger the proportional gain, the smaller the static error. However, a too large proportional gain might cause actuator saturation. Since this is an error-based control action, it is impossible to eliminate the static error if a proportional controller is used.
- The integral action eliminates the static error for a step reference or disturbance. It provides a slow action in the control signal because it is proportional to the accumulation of the past control error. Due to the slowness, the integral action is not good for the stability of the closed-loop system and it is easy to cause a large overshoot. In order to reduce the overshoot, the integral gain needs to be reduced. In the case of actuator saturation, it also causes an undesirable effect known as wind-up and certain anti-wind-up techniques are needed [6].
- The derivative term provides a fast action according to the future trend of the control error. It counteracts the effect of the integral term to some extent and often improves the stability of the system. The overshoot decreases with increasing derivative time, but increases again when the

derivative time becomes too large. The ideal differentiator tends to yield a large control signal when there is a sudden change in the set-point, known as the set-point kick, and when there is a high-frequency control error such as that induced by measurement noises. In practice, an ideal differentiator is implemented with a cascaded low-pass filter to attenuate high-frequency noise (in theory, this is to make it proper). For a time-delay system, in particular when the delay is long in comparison to the time constant of the system, the derivative term does not help and is often switched off [2, 6, 7].

Ziegler–Nichols Method

The Ziegler–Nichols tuning method is a very well-known empirical PID tuning method. Although the resulting response is often oscillatory and there are many other better model-based tuning methods available nowadays, it is still worth looking at it. This method was proposed for the system for which a satisfactory model is in the form of a first-order-plus-dead-time (FOPDT)[1] given as

$$G(s) = \frac{K}{Ts+1}e^{-sh}, \qquad (2.2)$$

where K is the static gain of the plant, $T > 0$ is the time constant and h is the transparent delay or dead time. It is only valid for open-loop stable systems and the tuning is carried out in three steps:

(i) Set the plant under P control with a very small gain for a step reference.

(ii) Gradually increase the gain until the loop (more specifically, the control signal) starts oscillating. Record the corresponding gain K_u, known as the *ultimate gain*, and the oscillation period T_u, known as the *ultimate* period.

(iii) Set the control parameters of (2.1) according to Table 2.1.

Table 2.1. Ziegler–Nichols tuning formulas

Type of controller	K_P	T_i	T_d
P	$0.5K_u$		
PI	$0.45K_u$	$\frac{1}{1.2}T_u$	
PID	$0.6K_u$	$0.5T_u$	$0.125T_u$

As is well known that if the Nyquist plot of the above system crosses the critical point $(-1, 0)$, the system is critically stable. Hence, the above procedure is to find out the proportional gain K_u that makes the system critically stable and then determine the parameters. Apparently, one point

---
[1] The major properties of many chemical processes can be captured by this model [124].

is not enough to characterise the whole system. The resulting system has a damping ratio close to 0.2, which is too small and not satisfactory for many systems. Nevertheless, it offers a very good starting point for further fine tuning.

Another drawback is that it is difficult to apply to working plants. A sudden change in the control signal or operation at the critically stable condition is not acceptable for critical processes. In this case, the relay-feedback approach [5, 6, 145] can be applied to identify the plant parameters for tuning, even for auto-tuning.

Analytical Tuning Based on Gain and Phase Margins

A control system is often designed to meet specified gain and phase margins so that the system is robustly stable.

Here, for the time-delay plant $G(s)$ given in (2.2), consider the unity feedback system shown in Figure 2.1 with the PI controller

$$C(s) = K_p(1 + \frac{1}{T_i s}). \tag{2.3}$$

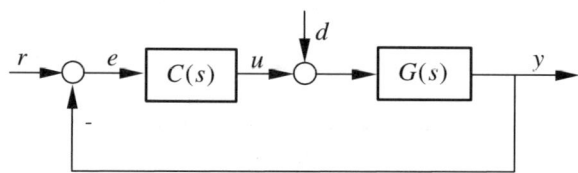

Figure 2.1. Unity negative feedback control system

The gain margin A_m of the system is defined as

$$A_m = \frac{1}{|C(j\omega_p)G(j\omega_p)|},$$

where the phase crossover frequency ω_p satisfies

$$\arg[C(j\omega_p)G(j\omega_p)] = -\pi.$$

The phase margin ϕ_m of the system is defined as

$$\phi_m = \arg[C(j\omega_g)G(j\omega_g)] + \pi,$$

where the gain crossover frequency ω_g satisfies

$$|C(j\omega_g)G(j\omega_g)| = 1.$$

For the given gain margin A_m and phase margin ϕ_m, the parameters of $C(s)$ are given by [46]

$$K_p = \frac{T\omega_p}{A_m K}, \quad T_i = (2\omega_p - \frac{4\omega_p^2 h}{\pi} + \frac{1}{T})^{-1},$$

where

$$\omega_p = \frac{A_m \phi_m + 0.5\pi A_m(A_m - 1)}{(A_m^2 - 1)h}.$$

Note that this does not offer the exact phase and gain margins, because the following approximation was used to derive the formula:

$$\arctan x = \begin{cases} \frac{\pi}{4}x & (|x| \leq 1), \\ \frac{\pi}{2} - \frac{\pi}{4x} & (|x| > 1). \end{cases}$$

2.1.3 Simulation Examples

Two examples are given here with one PI controller tuned using the Ziegler–Nichols method (denoted Z-N in figures and tables) and the other tuned to meet the given gain margin $A_m = 3$ and phase margin $\phi_m = 60°$ (denoted H-H-C in figures and tables).

Example 1

The plant under control is

$$G(s) = \frac{1}{s+1} e^{-0.1s}.$$

This is a lag-dominant FOPDT (the time constant is much larger than the delay). The parameters of PI controllers are given in Table 2.2 and the corresponding system responses are shown in Figure 2.2. A step disturbance $d = -0.5$ was applied at $t = 4$ s. The set-point response obtained by the Z-N method is more oscillatory than the other one, but the disturbance response is faster.

Table 2.2. PI control parameters for Example 1

	K_p	T_i
Z-N	7.38	0.32
H-H-C	5.24	1.0

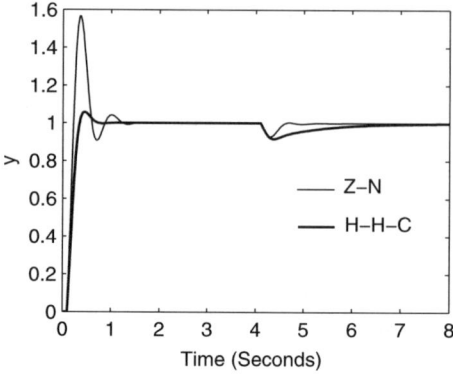

Figure 2.2. PID control: Example 1

Example 2

The plant under control is

$$G(s) = \frac{1}{s+1} e^{-2s}.$$

This is a delay-dominant FOPDT (the time constant $T = 1$ s while the delay $h = 2$ s). The parameters of PI controllers are given in Table 2.3 and the corresponding system responses are shown in Figure 2.3. A step disturbance $d = -0.5$ was applied at $t = 20$ s. The response obtained by the Z-N method is slow and undesirable, which indicates that the Z-N method is not good for delay-dominant processes. The H-H-C method works well.

Table 2.3. PI control parameters for Example 2

	K_p	T_i
Z-N	0.68	4.59
H-H-C	0.26	1.0

2.2 Smith Predictor (SP)-based Control

2.2.1 Control Difficulties Due to Delay

Although many processes can be controlled by PI(D) controllers, it does not mean that everything is perfect. Consider again the typical control system shown in Figure 2.1 with a PI controller (2.3) and an FOPDT plant (2.2).

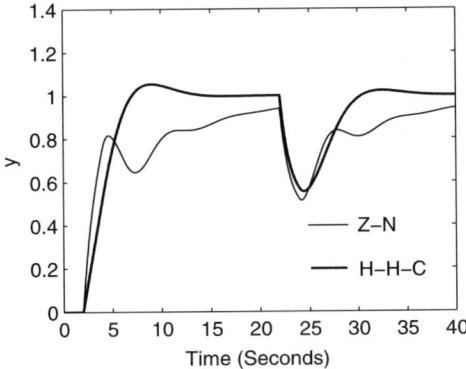

Figure 2.3. PID control: Example 2

The transfer function from the reference r to the output y is

$$T_{yr}(s) = \frac{C(s)G(s)}{1+C(s)G(s)} = \frac{KK_p(T_i s+1)e^{-sh}}{T_i s(Ts+1) + KK_p(T_i s+1)e^{-sh}}$$

and the characteristic equation of the closed-loop system is

$$T_i s(Ts+1) + KK_p(T_i s+1)e^{-sh} = 0.$$

This is a transcendental equation. In general, the delay term appears in the closed-loop characteristic equation and it is difficult to analyse the system stability or to design a controller to guarantee the stability.

Further analysis reveals that the delay also limits the maximum controller gain considerably, in particular, when the delay is relatively long. Assume that the PI controller is designed to cancel the lag,[2] *i.e.*, to choose $T_i = T$, then

$$T_{yr}(s) = \frac{KK_p e^{-sh}}{Ts + KK_p e^{-sh}}.$$

The system is stable only when

$$0 \leq KK_p < \frac{\pi}{2}\frac{T}{h}.$$

The controller gain is inversely proportional to the delay h. The longer the delay, the smaller the maximum allowable gain. This often means the performance is limited by the delay and a sluggish response is obtained.

---

[2] In this case, the plant pole $s = -\frac{1}{T}$, which disappears from the set-point response, remains in the disturbance response. See [84] for more details about whether to cancel or to shift a stable mode, which is a fundamental trade-off between (input) disturbance rejection and robustness to model errors and output disturbance rejection.

2.2.2 Smith Predictor

The Smith predictor, which was proposed in the late 1950s [126], aims to design a controller for a time-delay system such that it results in a delayed response of a delay-free system, as if the delay were shifted outside the feedback loop. Hence, the control design and system analysis are considerably simplified. This is realized by introducing local feedback to the main controller $C(s)$ using the Smith predictor $Z(s)$, as shown in Figure 2.4. Here, the plant $G(s) = P(s)e^{-sh}$ is assumed to be stable and the Smith predictor

$$Z(s) = P(s) - P(s)e^{-sh} \qquad (2.4)$$

is implemented using models of the delay-free part $P(s)$ of the plant, and the plant $P(s)e^{-sh}$.

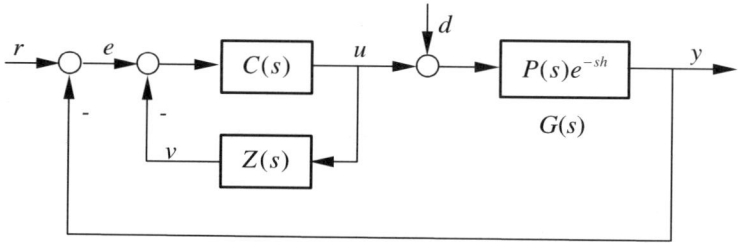

Figure 2.4. SP-based control system

Assume that $d = 0$ and there is no modelling error, then

$$\begin{aligned} y + v &= G(s)u(s) + Z(s)u(s) \\ &= P(s)u(s) \\ &= P(s)e^{-sh}u(s) \cdot e^{sh} \\ &= y \cdot e^{sh}. \end{aligned}$$

The feedback signal for the main controller $C(s)$ is a predicted version of y. This explains why it is called a predictor. When $P(s)$ is stable, the system shown in Figure 2.4 is equivalent to the one shown in Figure 2.5, which is the internal model control (IMC) version. When the model is exactly the same as the plant[3] and the disturbance is assumed to be $d = 0$, the signal y_0 is 0 and the outer loop can be regarded as open. In this case, the system is equivalently shown in Figure 2.6. It can be seen that the delay is moved outside the feedback loop and the main controller $C(s)$ can be designed according to the delay-free part $P(s)$ of the plant only. The aforementioned gain constraint on the

---

[3] The notation of the model and the plant are the same to reflect the nominal case. This should not cause any confusion.

controller $C(s)$ no longer exists, at least, explicitly. However, this does not mean that the controller gain can be excessively high because the delay still puts fundamental limitations on the achievable bandwidth [4]. The controller gain still has to be a compromise between the robustness and the speed of the system.

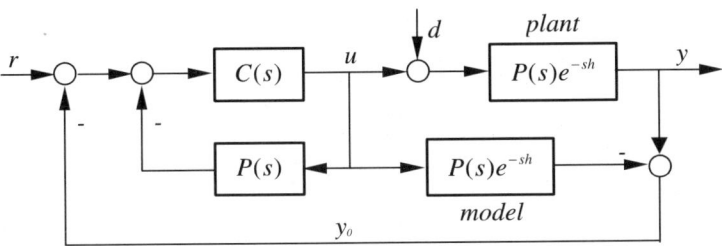

Figure 2.5. SP-based control system: Internal model control

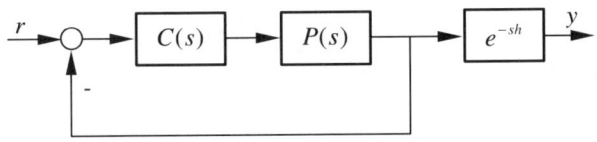

Figure 2.6. SP-based control system: Nominal case and $d = 0$

2.2.3 Robustness

In the nominal case, the controller for a stable system $G(s) = P(s)e^{-sh}$ can be designed in two steps:

(i) design a controller $C(s)$ for $P(s)$ to meet the given specifications;
(ii) incorporate a local feedback loop using the Smith predictor $Z = P - Pe^{-sh}$ to construct the controller for the delay system.

However, the world is not perfect. Sometimes, it is impossible to find the exact model of the plant. In this case, the predictor $Z(s)$ has to be constructed using the model $\bar{P}(s)e^{-s\bar{h}}$ of the plant as

$$Z(s) = \bar{P}(s) - \bar{P}(s)e^{-s\bar{h}}.$$

The modelling error can be represented by the multiplicative uncertainty $\Delta(s)$ via

$$P(s)e^{-sh} = \bar{P}(s)e^{-s\bar{h}}(1 + \Delta(s)).$$

Then, the delay term will not disappear from the closed-loop characteristic equation and the transfer function from r to y in Figure 2.4 is

$$\begin{aligned} T_{yr} &= \frac{CPe^{-sh}}{1 + C\bar{P} - C\bar{P}e^{-s\bar{h}} + CPe^{-sh}} \\ &= \frac{C\bar{P}e^{-s\bar{h}}(1 + \Delta)}{1 + C\bar{P} + C\bar{P}e^{-s\bar{h}}\Delta} \\ &= \frac{T_{yr}^0(1 + \Delta)}{1 + T_{yr}^0 \Delta e^{-s\bar{h}}} e^{-s\bar{h}}, \end{aligned}$$

where

$$T_{yr}^0 = \frac{C\bar{P}}{1 + C\bar{P}}$$

is the closed-loop transfer function of the delay-free system and is designed to be stable. For any stable multiplicative uncertainty $\Delta(s)$, according to the well-known small-gain theorem [40, 180], the closed-loop system is robustly stable if

$$\left|T_{yr}^0(j\omega)\Delta(j\omega)\right| < 1, \qquad (\forall \omega \geq 0).$$

In particular, if the modelling error exists in the delay term only, i.e., $\Delta(s) = e^{-s(h-\bar{h})} - 1$, then the above condition becomes

$$\left|T_{yr}^0(j\omega)\right| < \frac{1}{2}, \qquad (\forall \omega \geq 0).$$

This condition is conservative, but it guarantees stability for any delay mismatch.

The robustness with respect to mismatches in the delay h has been extensively investigated in the literature for different systems; see e.g., [34, 65, 112]. This is called practical stability in [112, 113] and the closed-loop system discussed above is practically stable for infinitesimal delay mismatches if and only if

$$\lim_{\omega \to \infty} \left|T_{yr}^0(j\omega)\right| < \frac{1}{2}. \tag{2.5}$$

A more general notion is the w-stability [34]. For the system discussed here, it is w-stable (for infinitesimal delay mismatches) if and only if C and $\bar{P}$ are proper and

$$\left|C(\infty)\bar{P}(\infty)\right| < 1,$$

in addition to C stabilising $\bar{P}$. This is consistent with (2.5). If either $C(s)$ or $\bar{P}(s)$ is strictly proper and the other is proper, then the w-stability problem does not occur.

2.2.4 Disturbance Rejection

So far, only the tracking performance has been studied. In this subsection, the regulatory performance (disturbance rejection) is discussed.

Assume that $r = 0$ and there is no modelling error, then according to Figure 2.4, the transfer function from d to y is

$$\begin{aligned} T_{yd} &= \frac{Pe^{-sh}}{1 + \frac{C}{1+CZ}Pe^{-sh}} \\ &= \frac{P(1 + CZ)e^{-sh}}{1 + CP} \\ &= \frac{CP}{1 + CP}Ze^{-sh} + \frac{P}{1 + CP}e^{-sh}. \end{aligned} \qquad (2.6)$$

It consists of two parts. The poles of the disturbance response include those of the closed-loop set-point response and those of the predictor Z. For the classical SP considered here, the poles of Z are those of the open-loop plant. In other words, the open-loop plant poles appear in the disturbance response.[4] Since C is designed to stabilise P, T_{yd} is stable if and only if Z is stable. This is true for $Z = P - Pe^{-sh}$ when P is stable. However, if P is unstable, then T_{yd} is not stable and hence the system is not stable. This means that the (classical) SP is only applicable to stable plants. If it is possible to find a predictor Z such that it is stable for an unstable plant, then the Smith predictor can still be used. This motivates the modified Smith predictor discussed in the next section.

It is easy to design a controller C for P to provide zero static error for a step reference r: simply guarantee that there is at least one integrator in CP. Since the static gain of Z is 0, in order to obtain zero static error for a step disturbance, it is necessary that

$$\lim_{s \to 0} \frac{P}{1 + CP} = 0.$$

This means that the integrator in CP should be in C. When there is an integrator in C and CP, there is no static error in the disturbance response even if there exist model mismatches, provided that the system is stable.

In summary, when designing an SP-based controller for a time-delay system, the following conditions should be met:
 (i) C stabilises $\bar{P}$,
 (ii) $|C(\infty)\bar{P}(\infty)| < 1$,
 (iii) there exists an integrator in C and CP (for step references and disturbances).

---

[4] This can be changed by using another predictor, e.g., the modified Smith predictor discussed later.

2.2.5 Simulation Examples

Example 1

Consider the plant $G(s) = \frac{1}{s+1}e^{-2s}$ studied in Subsection 2.1.3. Here, the controller incorporates a main controller $C(s) = K_p(1 + \frac{1}{T_i s})$ and the SP

$$Z(s) = \frac{1}{s+1}(1 - e^{-2s}).$$

The integral time T_i is chosen to be equal to the lag, i.e., $T_i = 1$, which results in a first-order set-point response with time constant $\frac{1}{K_p}$. All the conditions mentioned in the previous subsection are met and hence the system is stable and, moreover, robustly stable with respect to infinitesimal delay mismatches.

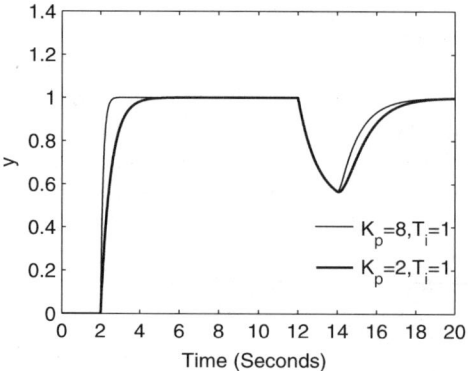

Figure 2.7. SP Example 1: Nominal responses

The nominal system responses are shown in Figure 2.7 for $K_p = 8$ and $K_p = 2$. A step disturbance $d = -0.5$ was applied at $t = 10$ s. The set-point response can be made as fast as possible, but the disturbance response is dominated by the open-loop dynamics and cannot be improved much (unless the predictor is changed; see the next section). As expected, the fast set-point response is obtained at the cost of robustness, as can be seen from the responses with different modelling errors shown in Figure 2.8. When $K_p = 8$, the system becomes unstable for a 10% mismatch in the delay. The system is very sensitive to delay mismatches, which was often blamed on the Smith predictor. However, it is in fact due to the aggressive requirement of the speed of the set-point response. It is recommended that the bandwidth of a system with delay h does not exceed $\frac{\pi}{2h}$ or $\frac{2}{h}$ [4]. Another way to improve the set-point response while maintaining robustness is to adopt the two-degree-of-freedom structure [47, 99], which considerably relaxes the fundamental limitations on the tracking performance [17, 128].

2.2 Smith Predictor (SP)-based Control

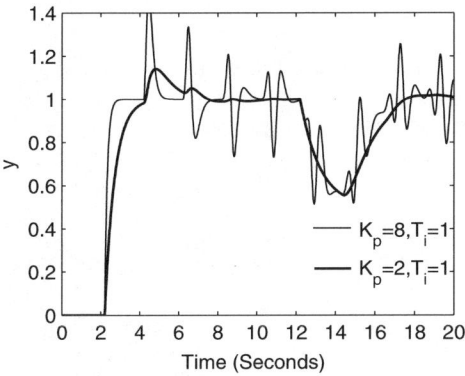

(a) h increased by 10%

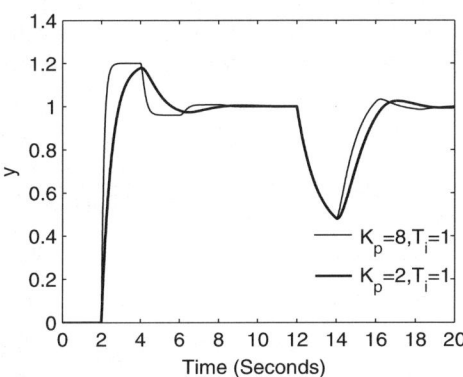

(b) K increased by 20%

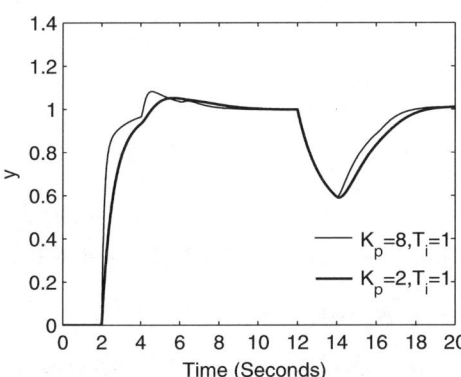

(c) T increased by 20%

Figure 2.8. SP Example 1: Robustness

Example 2

This example shows how an unstable delay system behaves when it is under the control of the SP scheme.

Consider the unstable system

$$G(s) = \frac{1}{s-1}e^{-2s}$$

with Smith predictor Z and main controller C given by

$$Z(s) = \frac{1}{s-1}(1 - e^{-2s}), \qquad C(s) = K_p(1 + \frac{1}{T_i s}).$$

In order to stabilise the delay-free system, it is necessary that $K_p > 1$. Here, the parameters are chosen as $K_p = 8$, $T_i = 1$. The system response is shown in Figure 2.9, where a step disturbance $d = -0.5$ was applied at $t = 16$ s. The set-point response seems stable, but it is very easy to show instability when different numerical solvers are used. The disturbance response is unstable.

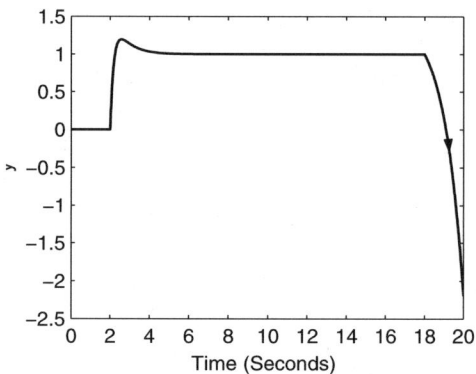

Figure 2.9. SP Example 2: Unstable plant

2.3 Modified Smith Predictor (MSP)-based Control

2.3.1 Modified Smith Predictor

As mentioned in the previous section, the classical Smith predictor cannot be applied to unstable time-delay systems. It is not a problem to design a controller C to stabilise the delay-free part of the system; the problem is that the predictor Z defined in (2.4) is unstable. If it is possible to find a stable Z, then the Smith predictor may still be of use.

2.3 Modified Smith Predictor (MSP)-based Control

Assume that the delay-free part P has a state-space realization

$$P = \left[\begin{array}{c|c} A & B \\ \hline C & 0 \end{array}\right]$$

and denote

$$\hat{P} = \left[\begin{array}{c|c} A & B \\ \hline Ce^{-Ah} & 0 \end{array}\right],$$

then

$$Z = \hat{P} - Pe^{-sh} = Ce^{-Ah}\int_0^h e^{-(sI-A)\zeta}d\zeta B \qquad (2.7)$$

is stable. This is known as the modified Smith predictor [147]. If this Z is plugged into Figure 2.4, then the scheme is applicable to unstable systems provided that Z is implemented as one stable block. Note that there might be unstable pole–zero cancellations inside Z if P is unstable. Because of this, Z cannot be implemented as the difference of $\hat{P}$ and Pe^{-sh} and its implementation is not trivial; see Part II for more details.

Plug (2.7) into Figure 2.4, assuming there is no modelling error, then

$$T_{yd} = \frac{Pe^{-sh}}{1 + \frac{C}{1+CZ}Pe^{-sh}} = \frac{P(1+CZ)e^{-sh}}{1 + CZ + CPe^{-sh}}$$
$$= \frac{CP}{1+C\hat{P}}Ze^{-sh} + \frac{P}{1+C\hat{P}}e^{-sh},$$

and

$$T_{yr} = \frac{\frac{C}{1+CZ}Pe^{-sh}}{1 + \frac{C}{1+CZ}Pe^{-sh}}$$
$$= \frac{CP}{1+C\hat{P}}e^{-sh}.$$

These equations are very similar to those obtained with the classical SP. What's different is that C should be designed to stabilise $\hat{P}$ so that $\frac{CP}{1+C\hat{P}}$ satisfies the given specification.

The MSP can be applied to some stable systems as well, although with some caution. See Chapter 10. In this case, the open-loop poles, which dominate the disturbance response in the SP case, can be removed and the MSP can be implemented as the difference of $\hat{P}$ and Pe^{-sh}.

2.3.2 Zero Static Error

Since, in general, $P(0) \neq \hat{P}(0)$, incorporating an integrator in $C(s)$ does not guarantee zero static error for step reference tracking or disturbance rejection. In order to recover this property, it is necessary to guarantee that the predictor Z has a zero static gain. Adding the constant $-C\int_0^h e^{-A\zeta}d\zeta B$ to $\hat{P}$ results in

32   2 Classical Control of Time-delay Systems

$$\hat{P} = \left[\begin{array}{c|c} A & B \\ \hline Ce^{-Ah} & 0 \end{array} \right] - C \int_0^h e^{-A\zeta} d\zeta B, \tag{2.8}$$

$$Z = Ce^{-Ah} \int_0^h e^{-(sI-A)\zeta} d\zeta B - C \int_0^h e^{-A\zeta} d\zeta B. \tag{2.9}$$

The above $\hat{P}$ and Z guarantees $\hat{P}(0) = P(0)$ and $Z(0) = 0$ and the static error is zero for step references and disturbances.

2.3.3 Simulation Examples

Two examples are given: one is a stable system to show that the dominant dynamics in the disturbance response can be removed, and the other is an unstable system to show the effectiveness of MSP.

Example 1

Consider the plant $G(s) = \frac{1}{s+1} e^{-0.1s}$ studied in Subsection 2.1.3. Here, the controller in Figure 2.4 incorporates the MSP

$$Z = \frac{e^{0.1}}{s+1} - (e^{0.1} - 1) - \frac{1}{s+1} e^{-0.1s}.$$

The main controller is chosen to be $C(s) = K_p(1 + \frac{1}{T_i s})$ to stabilise

$$\hat{P} = \frac{e^{0.1}}{s+1} - (e^{0.1} - 1) = \frac{(1 - e^{0.1})s + 1}{s+1}.$$

As a result,

$$\frac{CP}{1+C\hat{P}} = \frac{K_p(T_i s + 1)}{T_i s(s+1) + K_p(T_i s + 1)((1 - e^{0.1})s + 1)}. \tag{2.10}$$

From this, the stability condition (for positive parameters) is found to be

$$0 < K_p < \frac{1}{e^{0.1} - 1}, \quad T_i > \frac{K_p}{K_p + 1}(e^{0.1} - 1).$$

This is shown as the shaded area in Figure 2.10.

Case 1: $T_i \geq 1$

The root locus of the system $\frac{K_p}{T_i} \frac{T_i s + 1}{s} \frac{(1-e^{0.1})s+1}{s+1}$ has the form shown in Figure 2.11(a). The zero $s = -1/T_i$ lies between $s = 0$ and $s = -1$. Hence, the two closed-loop poles are all real and one of them is between 0 and 1. This pole dominates the system response and, hence, the closed-loop response is slower than the open-loop response. This can be verified from the system responses shown in Figure 2.12 for $K_p = 7$ and $T_i = 2$, where a step disturbance $d = -0.5$

2.3 Modified Smith Predictor (MSP)-based Control

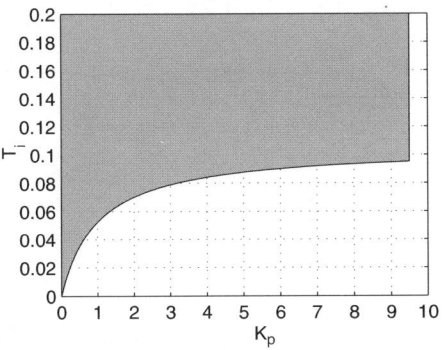

Figure 2.10. MSP Example 1: Stability region of (K_p, T_i)

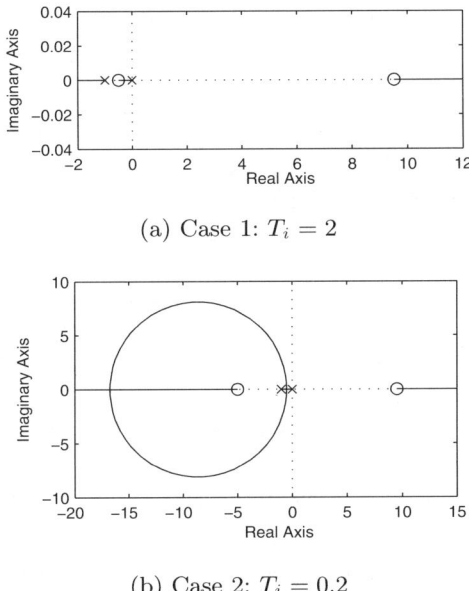

(a) Case 1: $T_i = 2$

(b) Case 2: $T_i = 0.2$

Figure 2.11. Root-locus design of PI controller for MSP-based stable system

was applied at $t = 5$ s. The dominant time constant is about 2.1 s. Although robustness is very good, the system response is too slow.

Case 2: $\frac{K_p}{K_p+1}(e^{0.1} - 1) < T_i < 1$

In this case, the root locus of the system $\frac{K_p}{T_i} \frac{T_i s+1}{s} \frac{(1-e^{0.1})s+1}{s+1}$ has the form shown in Figure 2.11(b). The zero $s = -1/T_i$ lies on the left of the pole $s = -1$.

34   2 Classical Control of Time-delay Systems

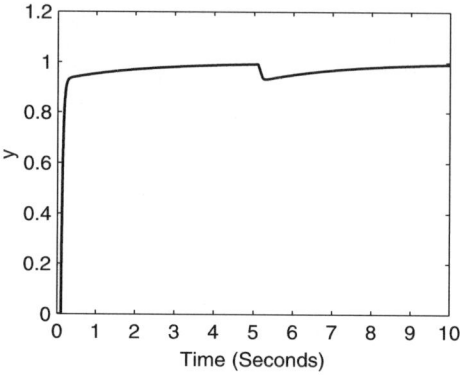

(a) Nominal response

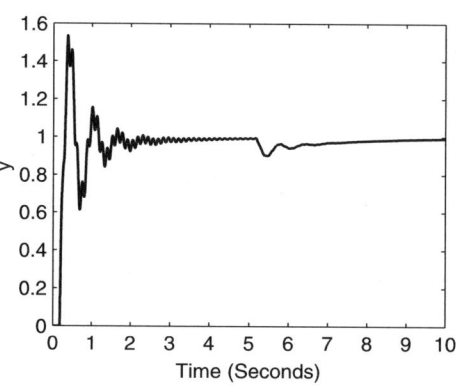

(b) Robust response when h increases by 80%

Figure 2.12. Performance of the MSP-based stable system: $K_p = 7$, $T_i = 2$

Bearing in mind that the closed-loop system bandwidth is limited to around $\frac{\pi}{2h} = 15.7$ rad/s, the control parameters are then chosen as $K_p = 7$, $T_i = 0.2$ to provide a damping ratio of 0.709 and a pair of complex closed-loop poles $-8.15 \pm j8.1$. The system responses are shown in Figure 2.13, where a step disturbance $d = -0.5$ was applied at $t = 5$ s. It can be seen that the system response is very fast and the open-loop dynamics (with a time constant of 1) no longer exists in the system response. The set-point response seems undesirable, but the large overshoot is due to the term $T_i s + 1$ in the numerator of the response (2.10). The overshoot can be considerably reduced by adding a low-pass filter $\frac{1}{T_i s + 1}$ to the set-point, which offers a 2-degree-of-freedom control structure. See Figure 2.13(c) for the corresponding nominal response. The 5% overshoot is now consistent with the damping ratio of 0.709.

2.3 Modified Smith Predictor (MSP)-based Control

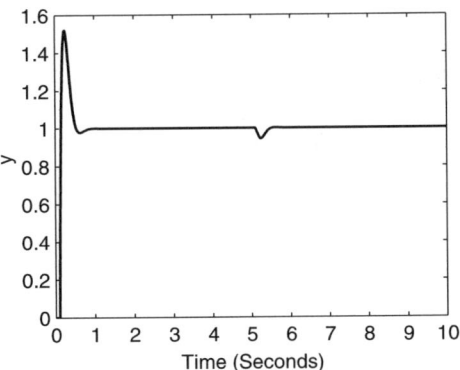

(a) Nominal response

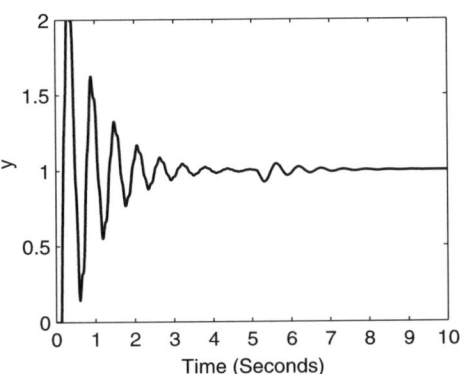

(b) Robust response when h increases by 40%

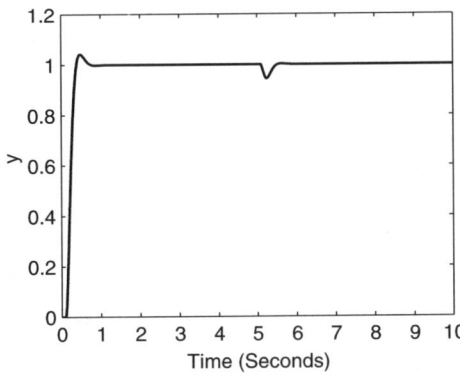

(c) Nominal response with a low-pass filter

Figure 2.13. Performance of the MSP-based stable system: $K_p = 7$, $T_i = 0.2$

Example 2

Consider the following unstable system:

$$G(s) = \frac{1}{s-1} e^{-sh}.$$

According to (2.8, 2.9), then

$$\hat{P} = \frac{e^{-h}}{s-1} + e^{-h} - 1 = \frac{(e^{-h}-1)s+1}{s-1},$$

$$Z = \frac{e^{-h}}{s-1} + e^{-h} - 1 - \frac{1}{s-1} e^{-sh}.$$

The main controller is chosen to be $C(s) = K_p(1 + \frac{1}{T_i s})$. As a result,

$$\frac{CP}{1+C\hat{P}} = \frac{K_p(T_i s + 1)}{T_i s(s-1) + K_p(T_i s + 1)((e^{-h}-1)s+1)}. \quad (2.11)$$

It can be found that the stability condition (for positive parameters) is[5]

$$1 < K_p < \frac{1}{1-e^{-h}}, \quad T_i > \frac{K_p}{K_p - 1}(1 - e^{-h}).$$

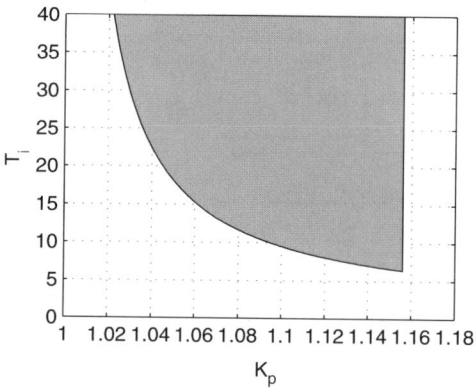

Figure 2.14. MSP Example 2: Stability region of (K_p, T_i)

This is the shaded area shown in Figure 2.14 for $h = 2$ (the controller will be designed for $h = 2$ s). The admissible gain has a very limited range: from 1

---

[5] The first condition also guarantees $\left|C(\infty)\hat{P}(\infty)\right| < 1$.

to $\frac{1}{1-e^{-h}}$. The longer the delay h, the narrower the gain range. Moreover, the integral time has to be large enough as well. The longer the delay h, the larger the integral time T_i. This is because the delay and the unstable pole all impose fundamental limitations on the achievable performances [4, 17, 116, 128].

It is very tricky when tuning the parameters. Only when T_i is large enough, it is possible to find a stabilising controller. For $T_i = 50$, the root locus is shown in Figure 2.15, from which K_p is chosen as $K_p = 1.08$ to obtain a damping ratio of 0.828 and a pair of complex closed-loop poles $-0.475 \pm j0.322$. This offers an overshoot of about 1%.

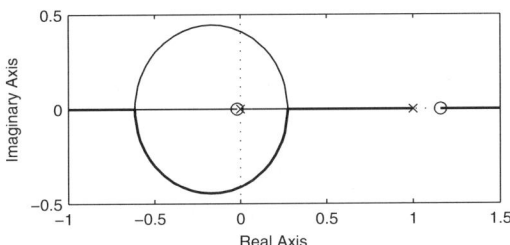

Figure 2.15. Root-locus design for MSP-based unstable system

The system responses are shown in Figure 2.16, where a step disturbance $d = -0.1$ was applied at $t = 20$ s and a filter $\frac{1}{T_i s+1}$ was added to the set-point. Figure 2.16(a) shows the nominal responses: the unstable one, denoted as "difference", obtained when Z is implemented as the difference of $\hat{P}$ and Pe^{-sh} and the stable one obtained when Z is implemented as

$$Z = e^{-h} - 1 + e^{-h} \cdot \frac{e^{\frac{h}{N}\epsilon} - e^{-\frac{h}{N}s}}{e^{\frac{h}{N}\epsilon} - 1} \cdot \frac{e^{\frac{h}{N}} - 1}{s/\epsilon + 1} \cdot \sum_{i=0}^{N-1} e^{-i\frac{h}{N}(s-1)},$$

with $N = 40$ and $\epsilon = 0.01$. See Chapter 11 for more details about this implementation. There is a very big dynamic error in the disturbance response. This is because the unstable pole works in open-loop for a period of $h = 2$ s, only after which the feedback controller starts regulating the disturbance. So the minimal maximum dynamic error is

$$d \cdot e^h = -0.1 \times e^2 = -0.74.$$

If the control parameters are chosen as $K_p = 1.112$ and $T_i = 20$, then the system is faster and the maximum dynamic error is close to the minimum one -0.74; see Figure 2.16(b). However, the system is not robustly stable even for $h = 2.05$ s. The slower one with $T_i = 50$ is robustly stable for $h = 2.05$ s; see Figure 2.16(c). The system is very easy to destabilise by modelling errors, because of the long delay and the unstable pole.

2 Classical Control of Time-delay Systems

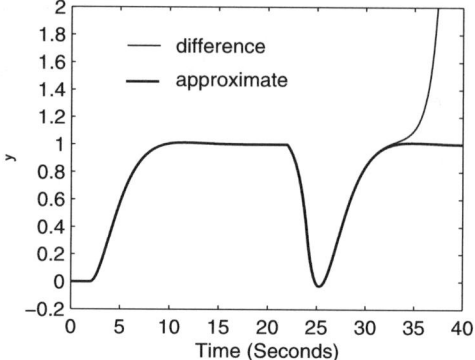

(a) Nominal response: different implementations

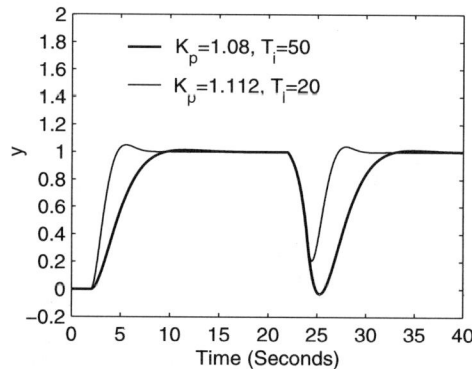

(a) Nominal response: different parameters

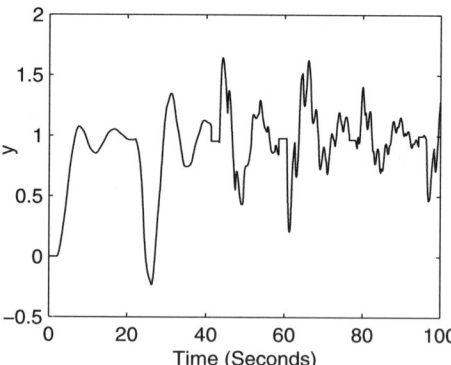

(b) Robust response: h increases to $h = 2.05$ s ($K_p = 1.08$, $T_i = 50$)

Figure 2.16. Performance of the MSP-based unstable system

2.4 Finite-spectrum Assignment (FSA)

The SP-based control scheme is very effective for control of stable time-delay systems. While the modified Smith predictor was being developed for unstable systems, another effective control strategy, known as finite-spectrum assignment nowadays, was developed for unstable systems as well. This strategy can address delays not only in the input/output channel, but also in the states. The delays can be multiple, commensurate and even distributed. Here, only the case with a single input delay is discussed, as the book focuses on systems with a single input/output delay. See [70, 108, 109, 146, 148, 149] for other cases and [49, 144] for a different version using the pole-assignment argument.

Consider a system described in the state-space realization as

$$\dot{x}(t) = Ax(t) + Bu(t-h),$$
$$y(t) = Cx(t).$$

Then, the plant transfer function is

$$G(s) = P(s)e^{-sh} = \left[\begin{array}{c|c} A & B \\ \hline C & 0 \end{array}\right] e^{-sh}.$$

The FSA adopts a state feedback control law

$$u(t) = Fx_p(t)$$

using the predicted state $x_p(t)$ given by

$$x_p(t) = e^{Ah}x(t) + \int_0^h e^{A\zeta} Bu(t-\zeta)d\zeta.$$

The resulting closed-loop system, having a finite spectrum, is stable if $A+BF$ is stable. Similar to the predictor-based control schemes, the delay term is removed from the design process.

This strategy will be discussed further with comparison to the modified Smith predictor in the next section. Simulation examples will be given in Part II with different implementations of the distributed delay

$$v(t) = \int_0^h e^{A\zeta} Bu(t-\zeta)d\zeta$$

in the control law.

2.5 Connection Between MSP and FSA

2.5.1 All Stabilising Controllers for Delay Systems

Assume that the control plant is $P(s)e^{-sh}$, of which the delay-free part P has the following realization:

$$P = \left[\begin{array}{c|c} A & B \\ \hline C & 0 \end{array}\right],$$

where (A, B) is stabilisable and (C, A) is detectable. Let F and L be such that $A + BF$ and $A + LC$ are Hurwitz. Then all stabilising controllers [90, 91] incorporating a modified Smith predictor

$$Z = Ce^{-Ah}(I - e^{-(sI-A)h}) \cdot \left[\begin{array}{c|c} A & B \\ \hline I & 0 \end{array}\right], \tag{2.12}$$

can be parameterised as shown in Figure 2.17, where

$$J(s) = \left[\begin{array}{c|cc} A + BF + e^{Ah}LCe^{-Ah} & -e^{Ah}L & B \\ \hline F & 0 & I \\ -Ce^{-Ah} & I & 0 \end{array}\right]$$

and $Q(s)$ is arbitrary but stable. This controller parameterisation involves one-degree-of-freedom. Another parameterisation involving two-degrees-of-freedom can be found in Chapter 9.

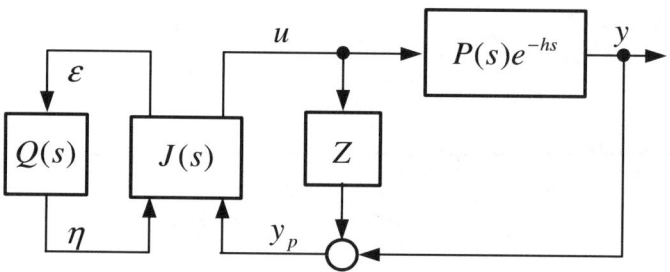

Figure 2.17. Stabilising controllers for processes with dead time

2.5.2 Predictor–Observer Representation: MSP

Denote the state vector of $J(s)$ by x_J, then

$$u = Fx_J + \eta, \qquad \varepsilon = -Ce^{-Ah}x_J + y_p.$$

The state equation of $J(s)$ is given by

$$\dot{x}_J = (A + BF + e^{Ah}LCe^{-Ah})x_J - e^{Ah}Ly_p + B\eta,$$

or equivalently, by

$$\begin{aligned} e^{-Ah}\dot{x}_J &= (A + LC)e^{-Ah}x_J - Ly_p + e^{-Ah}B(Fx_J + \eta) \\ &= (A + LC)e^{-Ah}x_J - Ly_p + e^{-Ah}Bu. \end{aligned}$$

Using the above formulae, all the stabilising controllers shown in Figure 2.17 can be represented as shown in Figure 2.18. It consists of an output predictor Z, a state observer and a state feedback. The state observer actually observes the states of the delay-free system $\left[\begin{array}{c|c}A & B \\ \hline I & 0\end{array}\right]$ because, in the nominal case,

$$
\begin{aligned}
x_J &= \left[\begin{array}{c|c}A+LC & e^{-Ah}B \\ \hline e^{Ah} & 0\end{array}\right] u - \left[\begin{array}{c|c}A+LC & L \\ \hline e^{Ah} & 0\end{array}\right](Z + Pe^{-sh})u \\
&= \left[\begin{array}{c|c}A+LC & e^{-Ah}B \\ \hline e^{Ah} & 0\end{array}\right] u - \left[\begin{array}{c|c}A+LC & L \\ \hline e^{Ah} & 0\end{array}\right] \left[\begin{array}{c|c}A & B \\ \hline Ce^{-Ah} & 0\end{array}\right] u \\
&= \left[\begin{array}{c|c}A+LC & I \\ \hline e^{Ah} & 0\end{array}\right] \left[\begin{array}{c|c}A & I \\ \hline -LC & I\end{array}\right] e^{-Ah} Bu \\
&= \left[\begin{array}{cc|c}A+LC & -LC & I \\ 0 & A & I \\ \hline e^{Ah} & 0 & 0\end{array}\right] e^{-Ah} Bu \\
&= \left[\begin{array}{c|c}A & B \\ \hline I & 0\end{array}\right] u.
\end{aligned}
$$

2.5.3 Observer–Predictor Representation: FSA

As mentioned in Section 2.4, the finite-spectrum assignment for time-delay systems is a state feedback control law

$$u(t) = Fx_p(t)$$

using the predicted state

$$x_p(t) = e^{Ah}x(t) + \int_0^h e^{A\zeta} Bu(t-\zeta)d\zeta.$$

If the state $x(t)$ is not available for prediction, then a Luenberger observer is needed to re-construct the state from the output y and the control u. It is easy to check that, in the nominal case,

$$
\begin{aligned}
x_o &= \left[\begin{array}{c|c}A+LC & I \\ \hline I & 0\end{array}\right] \cdot (Be^{-sh}u - Ly) \\
&= \left[\begin{array}{c|c}A+LC & I \\ \hline I & 0\end{array}\right] \cdot \left[\begin{array}{c|c}A & I \\ \hline -LC & I\end{array}\right] Be^{-sh}u \\
&= \left[\begin{array}{c|c}A & B \\ \hline I & 0\end{array}\right] e^{-sh} \cdot u
\end{aligned}
$$

gives the observed state x_o of the plant. Using these formulae, the FSA control structure (using output feedback via y) can then be depicted in Figure 2.19, assuming temporarily $Q(s) = 0$. It consists of a state observer and a state

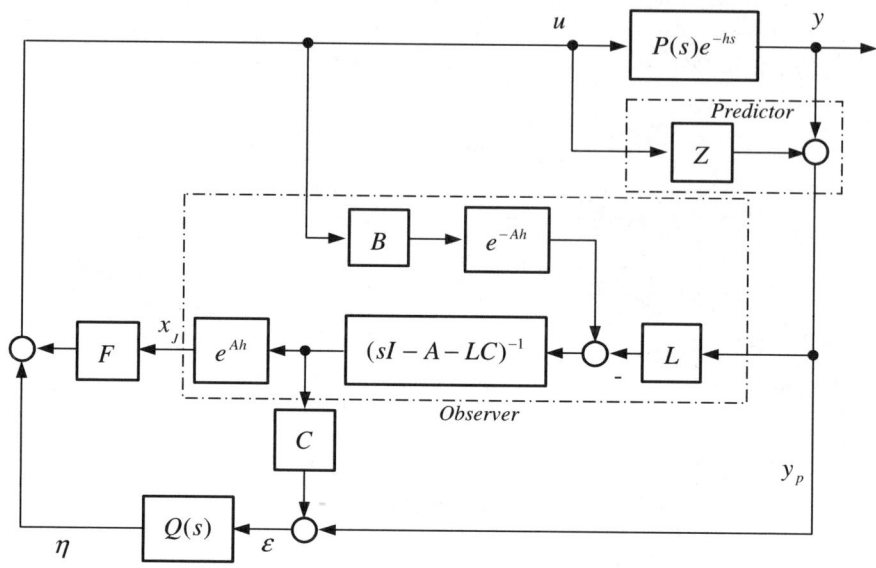

Figure 2.18. Predictor–Observer Representation: MSP scheme

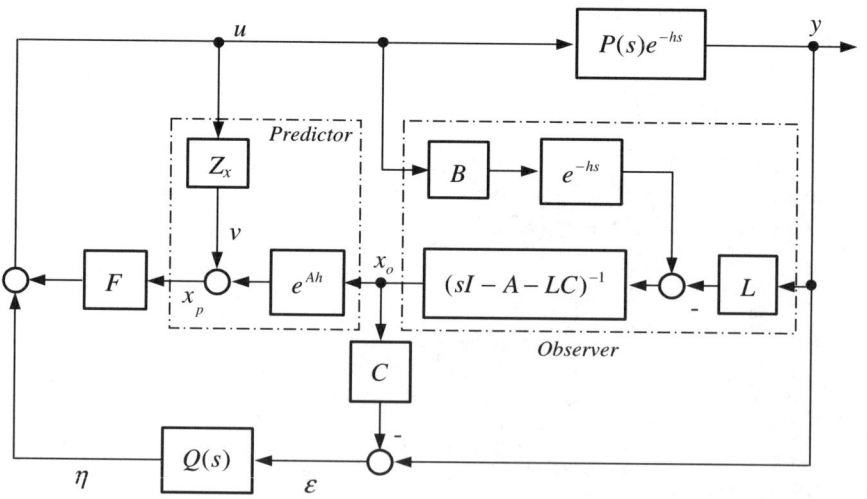

Figure 2.19. Observer–Predictor Representation: FSA scheme

predictor. As a matter of fact, this is exactly the central controller given in Figure 2.17, as described in [90, 91]. It is also easy to check that

$$\varepsilon = -Cx_o + y.$$

Hence, all the stabilising controller given in Figure 2.17 can also be represented as the observer-predictor structure shown in Figure 2.19, where the distributed delay

$$Z_x(s) = (I - e^{-(sI-A)h}) \cdot \left[\begin{array}{c|c} A & B \\ \hline I & 0 \end{array}\right]$$

is the transfer function from u to v of the block characterised by the integral

$$v(t) = \int_0^h e^{A\zeta} Bu(t-\zeta)d\zeta.$$

2.5.4 Some Remarks

Although the FSA scheme can deal with much more general time-delay systems, *e.g.*, systems with delays in the state and multiple delays, the FSA scheme and the MSP scheme are two equivalent representations of stabilising controllers for systems with a single input delay [91]. The two structures depicted in Figures 2.18 and 2.19 clearly show the similarities between them. Roughly speaking, only the order of the predictor and the observer is exchanged: in the FSA scheme, the observer goes first and then the predictor but, in the MSP scheme, the predictor goes first and then the observer. The only change from the FSA scheme to the MSP scheme is to change/move Z_x to Z and e^{-sh} in the observer to e^{-Ah}.

Some more insightful observations are:

(i) the predictor in the MSP scheme is an *output predictor* while that in the FSA scheme is a *state predictor*;

(ii) the observer in the MSP scheme is a state observer of the *delay-free* system while the one in the FSA scheme is a state observer of the *delay* system (and hence a state predictor has to be used before using the state feedback);

(iii) the free parameter $Q(s)$ may be used to improve the robustness of the FSA scheme w.r.t. the implementation error of Z_x;

(iv) in both cases, the central controller is essentially a state feedback controller, using the *predicted* state (x_p in the FSA scheme) or the *observed* state of the delay-free system (x_J in the MSP scheme).

3

Preliminaries

In this chapter, some preliminaries are collected for later use. These include two important FIR operators which map a rational transfer matrix into FIR blocks, the state-space operations of systems, the chain-scattering approach, algebraic Riccati equations, an important matrix called the Σ matrix, and the $L_2[0,h]$-induced norm.

3.1 FIR Operators

Finite-impulse-response (FIR) blocks (operators) play an important role in the control of dead-time systems [26, 70, 73, 80, 88, 105, 106, 115, 147, 159, 162]. Two such operators, the completion operator and the truncation operator are frequently used.

Assume that $G = \left[\begin{array}{c|c} A & B \\ \hline C & D \end{array}\right]$ is a rational transfer matrix, then for a given parameter $h \geq 0$, the *truncation operator* τ_h and the *completion operator* π_h are defined, respectively, as

$$\tau_h(G) \doteq \left[\begin{array}{c|c} A & B \\ \hline C & D \end{array}\right] - e^{-sh} \left[\begin{array}{c|c} A & e^{Ah}B \\ \hline C & 0 \end{array}\right] \doteq G - e^{-sh}\tilde{G},$$

$$\pi_h(G) \doteq \left[\begin{array}{c|c} A & B \\ \hline Ce^{-Ah} & 0 \end{array}\right] - e^{-sh} \left[\begin{array}{c|c} A & B \\ \hline C & D \end{array}\right] \doteq \hat{G} - e^{-sh}G.$$

This follows [85], except for a small adjustment in notation.

Note that these two operators map any rational transfer matrix G into an FIR block. The impulse response of $\tau_h(G)$ is the truncation of the impulse response of G to $[0,h]$. The impulse response of $\pi_h(G)$, which is also supported on $[0,h]$, is the only continuous function on $[0,h]$ with the following property: if it is added to the impulse response of $e^{-sh}G$, which is supported on $[h,\infty)$, then the impulse response of a rational transfer matrix, denoted above by $\hat{G}$, is obtained. As an example, the impulse responses of $\tau_h(G)$ and $\pi_h(G)$ are shown in Figure 3.1 for $G = \frac{1}{s-1}$ and $h = 1$.

46   3 Preliminaries

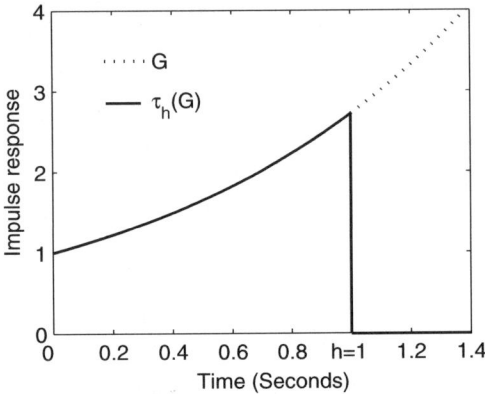

(a) the impulse responses of G and $\tau_h(G)$

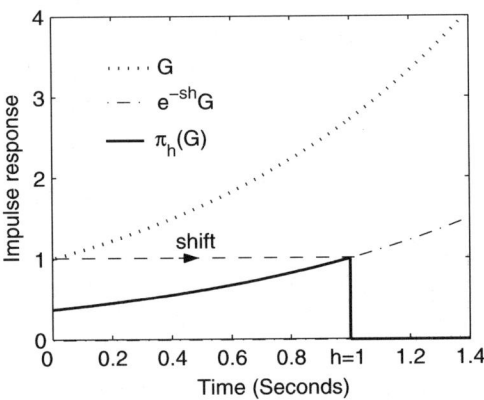

(b) the impulse responses of G, $e^{-sh}G$ and $\pi_h(G)$

Figure 3.1. Completion operator π_h and truncation operator τ_h

3.2 Chain-scattering Approach

This section describes the fundamentals of the chain-scattering approach from [58], but using slightly different notation as listed in Table 3.1.

3.2.1 Representations of a System: IOR and CSR

Consider the system
$$\begin{bmatrix} z \\ y \end{bmatrix} = M \begin{bmatrix} w \\ u \end{bmatrix}$$

3.2 Chain-scattering Approach

Table 3.1. CSR notation used in [58] and in this book

Notation used here	Notation used in [58]
$\mathcal{C}_r$	$CHAIN$
$\mathcal{C}_l$	$DCHAIN$
$\mathcal{H}_r$	HM
$\mathcal{H}_l$	DHM

shown in Figure 3.2, where $M = \begin{bmatrix} M_{11} & M_{12} \\ M_{21} & M_{22} \end{bmatrix}$. Since the exogenous disturbance w and the control u are all input signals and on one side of the block, and the controlled output z and the measurement y are all output signals and on the other side of the block, this representation is called the *input–output representation* (IOR) of the system M. This representation is widely used nowadays, in particular, in the context of robust control.

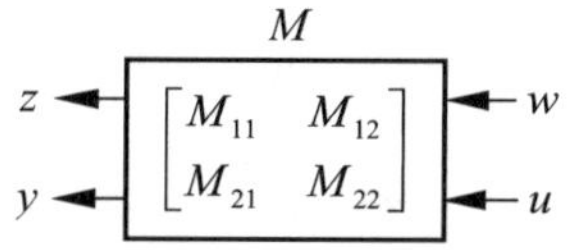

Figure 3.2. Input–output representation of a system

When M_{21}, *i.e.*, the transfer function from w to y, is invertible,

$$\mathcal{C}_r(M) \doteq \begin{bmatrix} M_{12} - M_{11}M_{21}^{-1}M_{22} & M_{11}M_{21}^{-1} \\ -M_{21}^{-1}M_{22} & M_{21}^{-1} \end{bmatrix}$$

is well defined and is called the *(right) chain-scattering transformation* (CST). Similarly, when M_{12}, *i.e.*, the transfer function from u to z, is invertible,

$$\mathcal{C}_l(M) \doteq \begin{bmatrix} M_{12}^{-1} & -M_{12}^{-1}M_{11} \\ M_{22}M_{12}^{-1} & M_{12} - M_{22}M_{12}^{-1}M_{11} \end{bmatrix}$$

is well defined and is called the (left) CST. If both M_{21} and M_{12} are invertible, then

$$\mathcal{C}_r(M) \cdot \mathcal{C}_l(M) = I.$$

In fact, $\mathcal{C}_l(M)$ satisfies

$$\begin{bmatrix} u \\ y \end{bmatrix} = \mathcal{C}_l(M) \begin{bmatrix} z \\ w \end{bmatrix},$$

which offers the *(left) chain-scattering representation* (CSR) for the system M, as shown in Figure 3.3(a), and $\mathcal{C}_r(M)$ satisfies

48 3 Preliminaries

$$\begin{bmatrix} z \\ w \end{bmatrix} = \mathcal{C}_r(M) \begin{bmatrix} u \\ y \end{bmatrix},$$

which offers the *(right) chain-scattering representation* (CSR) for the system M, as shown in Figure 3.3(b). The right or left CSR is just a different representation of the system M in the IOR shown in Figure 3.2 after swapping some signals, provided that M_{21} and M_{12} are, respectively, invertible. It is a simple idea that has brought many advantages to controller design [58] and is one of the foundations for this book.

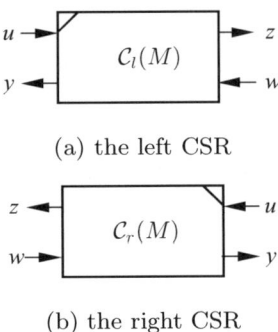

(a) the left CSR

(b) the right CSR

Figure 3.3. Chain-scattering representations of the system M in Figure 3.2

A tagged box is introduced to distinguish the left or right CSR (CST) in a block diagram. As shown in Figure 3.3, the tagged corner matches the *left* or *right* CST (or CSR) and also indicates the scattering direction. In other words, a tagged *right*-upper corner means that the matrix inside the box is the *right* CST of a system described in IOR, while a tagged *left*-upper corner means that the matrix inside the box is the *left* CST of a system described in IOR. This is very useful when the matrix inside the box is just a matrix without any indication of the right or left direction; see Figure 7.2 for an example.

The state-space realizations of $\mathcal{C}_r(M)$ and $\mathcal{C}_l(M)$ are derived in the following section.

3.2.2 Linear Fractional Transformations: The Standard LFT and the HMT

If the feedback controller for the system in Figure 3.2 is designed to be K, i.e.,

$$u = Ky, \tag{3.1}$$

then the closed-loop transfer function is

$$T_{zw} = \mathcal{F}_l(M, K) = M_{11} + M_{12}K(I - M_{22}K)^{-1}M_{21},$$

where $\mathcal{F}_l$ is the standard *lower linear fractional transformation* (LFT), provided that $I - M_{22}K$ is invertible. The dual, the *upper linear fractional transformation*, is defined as

$$\mathcal{F}_u(M, Q) = M_{22} + M_{21}Q(I - M_{11}Q)^{-1}M_{12}$$

provided that $I - M_{11}Q$ is invertible. The standard LFT $\mathcal{F}_l$ and $\mathcal{F}_u$ are associated with the IOR.

For the CSR, two less-used linear fractional transformations, called right and left *homographic transformations* (HMT) [22, 58] are defined for $N = \begin{bmatrix} N_{11} & N_{12} \\ N_{21} & N_{22} \end{bmatrix}$ and $\tilde{N} = \begin{bmatrix} \tilde{N}_{11} & \tilde{N}_{12} \\ \tilde{N}_{21} & \tilde{N}_{22} \end{bmatrix}$ as

$$\mathcal{H}_r(N, Q) = (N_{11}Q + N_{12})(N_{21}Q + N_{22})^{-1}$$

and

$$\mathcal{H}_l(\tilde{N}, Q) = -(\tilde{N}_{11} - Q\tilde{N}_{21})^{-1}(\tilde{N}_{12} - Q\tilde{N}_{22}),$$

where the dimensions of the matrices are compatible, provided that the corresponding inverse exists. The subscript l stands for *left* and r for *right*, indicating that the inverted term is at the *left* or *right* hand side. $\mathcal{H}_l$ and $\mathcal{H}_r$ are closely related to the *left* CST and the *right* CST, respectively. If $N = \mathcal{C}_r(M)$ and $\tilde{N} = \mathcal{C}_l(M)$, then the closed-loop transfer function, with the controller given in (3.1), is

$$T_{zw} = \mathcal{H}_r(N, K) = \mathcal{H}_r(\mathcal{C}_r(M), K)$$

for the right CSR shown in Figure 3.3(b), or

$$T_{zw} = \mathcal{H}_l(\tilde{N}, K) = \mathcal{H}_l(\mathcal{C}_l(M), K)$$

for the left CSR shown in Figure 3.3(a). It is now clear that the tagged port is also the port to be terminated by a "load" or controller K, or the port from which the system scatters.

3.2.3 Some Important Properties

The most important property of CSR is the cascade property: the cascade of matrices in the CSR is just multiplication of the matrices. For example,

$$\mathcal{H}_r(N_1, \mathcal{H}_r(N_2, Q)) = \mathcal{H}_r(N_1 N_2, Q)$$

where N_1 and N_2 are the right CSR of two systems, and

$$\mathcal{H}_l(\tilde{N}_1, \mathcal{H}_l(\tilde{N}_2, Q)) = \mathcal{H}_l(\tilde{N}_2 \tilde{N}_1, Q),$$

where $\tilde{N}_1$ and $\tilde{N}_2$ are the left CSR of two systems.

In some cases, the cascade property of the right and left HMT can considerably simplify the expositions, as will be seen later. It is wise to use LFT and HMT in parallel, when convenient. Some similarities between CSR and IOR are given in Table 3.2. For more details about the chain-scattering representation, see [58]. Only two lemmas are cited here for later use.

Table 3.2. Similarities between CSR and IOR

in the context of CSR	in the context of IOR
port-based, see Figure 3.3	input/output-based, see Figure 3.2
homographic transformation (HMT)	linear fractional transformation (LFT)
J-unitary	unitary
J-lossless	lossless or inner
J-lossless factorisation	inner-outer factorisation
J-spectral factorisation	spectral factorisation
cascade: direct multiplication	cascade: star product
termination	feedback

Lemma 3.1. [58, Lemma 4.13] $\mathcal{H}_r$ satisfies the following properties:
(i) $\mathcal{H}_r(\mathcal{C}_r(M), Q) = \mathcal{F}_l(M, Q)$;
(ii) $\mathcal{H}_r(I, Q) = Q$;
(iii) If $\mathcal{H}_r(G, Q) = R$ and G^{-1} exists, then $Q = \mathcal{H}_r(G^{-1}, R)$.

Lemma 3.2. [58, Lemma 7.1] Let Λ be any unimodular matrix, then the H^∞ control problem $\|\mathcal{H}_r(G, K_0)\|_\infty < \gamma$ is solvable iff $\|\mathcal{H}_r(G\Lambda, K)\|_\infty < \gamma$ is solvable. Furthermore, $K_0 = \mathcal{H}_r(\Lambda, K)$ or $K = \mathcal{H}_r(\Lambda^{-1}, K_0)$.

This lemma provides an approach to simplify an H^∞ control problem by *factoring* out a unimodular portion from the process, as used in [58]; it also paves the way to simplify an H^∞ control problem by *inserting* a unimodular portion to the process, as used in [162].

3.3 State-space Operations on Systems

State-space operations on systems are elementary tools in this book. Some operations on systems are summarised here.

In the rest of this section, assume that

$$G = \left[\begin{array}{c|c} A & B \\ \hline C & D \end{array}\right], \qquad G_1 = \left[\begin{array}{c|c} A_1 & B_1 \\ \hline C_1 & D_1 \end{array}\right], \qquad G_2 = \left[\begin{array}{c|c} A_2 & B_2 \\ \hline C_2 & D_2 \end{array}\right]$$

have appropriate dimensions for the operations to be carried out.

3.3.1 Operations on Systems

(i) Cascade/series connection or multiplication of G_1 and G_2

$$G_1 G_2 = \left[\begin{array}{c|c} A_1 & B_1 \\ \hline C_1 & D_1 \end{array}\right] \left[\begin{array}{c|c} A_2 & B_2 \\ \hline C_2 & D_2 \end{array}\right]$$

$$= \left[\begin{array}{cc|c} A_1 & B_1 C_2 & B_1 D_2 \\ 0 & A_2 & B_2 \\ \hline C_1 & D_1 C_2 & D_1 D_2 \end{array}\right] = \left[\begin{array}{cc|c} A_2 & 0 & B_2 \\ B_1 C_2 & A_1 & B_1 D_2 \\ \hline D_1 C_2 & C_1 & D_1 D_2 \end{array}\right].$$

(ii) Parallel connection or addition of G_1 and G_2

$$G_1 + G_2 = \left[\begin{array}{c|c} A_1 & B_1 \\ \hline C_1 & D_1 \end{array}\right] + \left[\begin{array}{c|c} A_2 & B_2 \\ \hline C_2 & D_2 \end{array}\right] = \left[\begin{array}{cc|c} A_1 & 0 & B_1 \\ 0 & A_2 & B_2 \\ \hline C_1 & C_2 & D_1 + D_2 \end{array}\right].$$

(iii) Inverse of G (when D is nonsingular)

$$G^{-1} = \left[\begin{array}{c|c} A - BD^{-1}C & -BD^{-1} \\ \hline D^{-1}C & D^{-1} \end{array}\right],$$

which means $G^{-1}G = GG^{-1} = I$.

(iv) Basic operations

$$G^T(s) = \left[\begin{array}{c|c} A^* & C^* \\ \hline B^* & D^* \end{array}\right],$$

$$G^\sim(s) = G^T(-s) = (G(-\bar{s}))^* = \left[\begin{array}{c|c} -A^* & -C^* \\ \hline B^* & D^* \end{array}\right],$$

$$\begin{bmatrix} G_1 & G_2 \end{bmatrix} = \left[\begin{array}{cc|cc} A_1 & 0 & B_1 & 0 \\ 0 & A_2 & 0 & B_2 \\ \hline C_1 & C_2 & D_1 & D_2 \end{array}\right],$$

$$\begin{bmatrix} G_1 \\ G_2 \end{bmatrix} = \left[\begin{array}{cc|c} A_1 & 0 & B_1 \\ 0 & A_2 & B_2 \\ \hline C_1 & 0 & D_1 \\ 0 & C_2 & D_2 \end{array}\right],$$

$$\begin{bmatrix} G_1 & 0 \\ 0 & G_2 \end{bmatrix} = \left[\begin{array}{cc|cc} A_1 & 0 & B_1 & 0 \\ 0 & A_2 & 0 & B_2 \\ \hline C_1 & 0 & D_1 & 0 \\ 0 & C_2 & 0 & D_2 \end{array}\right],$$

52   3 Preliminaries

$$G^\sim JG = \left[\begin{array}{cc|c} A & 0 & B \\ -C^*JC & -A^* & -C^*JD \\ \hline D^*JC & B^* & D^*JD \end{array}\right] = \left[\begin{array}{cc|c} -A^* & -C^*JC & -C^*JD \\ 0 & A & B \\ \hline B^* & D^*JC & D^*JD \end{array}\right],$$

$$GJG^\sim = \left[\begin{array}{cc|c} A & BJB^* & BJD^* \\ 0 & -A^* & -C^* \\ \hline C & DJB^* & DJD^* \end{array}\right] = \left[\begin{array}{cc|c} -A^* & 0 & -C^* \\ BJB^* & A & BJD^* \\ \hline DJB^* & C & DJD^* \end{array}\right].$$

(v) State feedback

If a state feedback $u = Fx + v$, as shown in Figure 3.4, is applied to G : $\begin{cases} \dot{x} = Ax + Bu \\ y = Cx + Du \end{cases}$, then the resulting system

$$G_{SF} : \begin{cases} \dot{x} = (A + BF)x + Bv \\ y = (C + DF)x + Dv \end{cases}$$

is realized as

$$G_{SF} = \left[\begin{array}{c|c} A & B \\ \hline C & D \end{array}\right] \left[\begin{array}{c|c} I & 0 \\ F & I \end{array}\right] = \left[\begin{array}{c|c} A + BF & B \\ \hline C + DF & D \end{array}\right],$$

where the multiplication is done as if the matrices were normal block matrices. This generates the right co-prime factorisation $G = NM^{-1}$ of G with

$$\left[\begin{array}{c} N \\ M \end{array}\right] = \left[\begin{array}{c|c} A & B \\ C & D \\ \hline 0 & I \end{array}\right] \left[\begin{array}{c|c} I & 0 \\ F & I \end{array}\right] = \left[\begin{array}{c|c} A + BF & B \\ C + DF & D \\ \hline F & I \end{array}\right].$$

On the other hand, if

$$\left[\begin{array}{c} N \\ M \end{array}\right] = \left[\begin{array}{c|c} A & B \\ L & K \\ \hline C & D \end{array}\right], \quad D \text{ is nonsingular}$$

then

$$NM^{-1} = \left[\begin{array}{c|c} A - BD^{-1}C & BD^{-1} \\ \hline L - KD^{-1}C & KD^{-1} \end{array}\right].$$

(vi) Output injection

If an output injection, as shown in Figure 3.5, is applied to the system G then the resulting system

$$G_{OI} : \begin{cases} \dot{x} = (A + LC)x + (B + LD)u \\ y = Cx + Du \end{cases}$$

is realized as

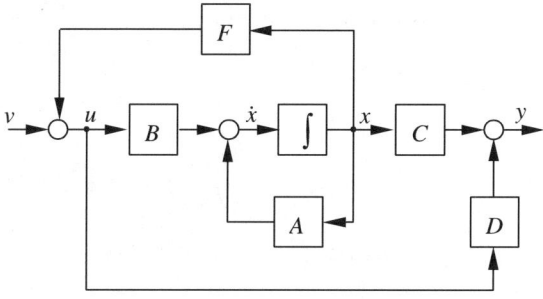

Figure 3.4. State feedback $u = Fx + v$

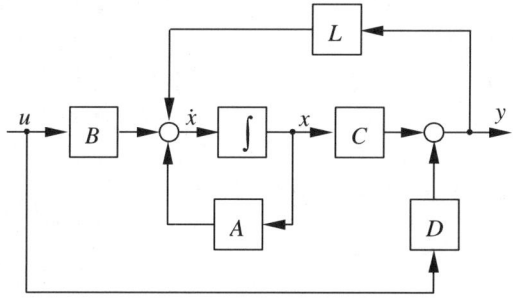

Figure 3.5. Output injection

$$G_{OI} = \begin{bmatrix} I & L \\ 0 & I \end{bmatrix} \left[\begin{array}{c|c} A & B \\ \hline C & D \end{array}\right] = \left[\begin{array}{c|c} A+LC & B+LD \\ \hline C & D \end{array}\right],$$

where the multiplication is done as if the two matrices were normal block matrices. This generates the left co-prime factorisation $G = \tilde{M}^{-1}\tilde{N}$ of G with

$$[\tilde{N} \; \tilde{M}] = \begin{bmatrix} I & L \\ 0 & I \end{bmatrix} \left[\begin{array}{c|cc} A & B & 0 \\ \hline C & D & I \end{array}\right] = \left[\begin{array}{c|cc} A+LC & B+LD & L \\ \hline C & D & I \end{array}\right].$$

On the other hand, if

$$[\tilde{N} \; \tilde{M}] = \left[\begin{array}{c|cc} A & L & B \\ \hline C & K & D \end{array}\right], \qquad D \text{ is nonsingular}$$

then

$$\tilde{M}^{-1}\tilde{N} = \left[\begin{array}{c|c} A - BD^{-1}C & L - BD^{-1}K \\ \hline D^{-1}C & D^{-1}K \end{array}\right].$$

(vii) Chain-scattering transformations

If the state-space realization of M is

$$M = \left[\begin{array}{c|cc} A & B_1 & B_2 \\ \hline C_1 & D_{11} & D_{12} \\ C_2 & D_{21} & D_{22} \end{array}\right],$$

then the state-space realizations of $\mathcal{C}_r(M)$ and $\mathcal{C}_l(M)$ are

$$\mathcal{C}_r(M) = \left[\begin{array}{c|ccc} A - B_1 D_{21}^{-1} C_2 & B_2 - B_1 D_{21}^{-1} D_{22} & B_1 D_{21}^{-1} \\ \hline C_1 - D_{11} D_{21}^{-1} C_2 & D_{12} - D_{11} D_{21}^{-1} D_{22} & D_{11} D_{21}^{-1} \\ -D_{21}^{-1} C_2 & -D_{21}^{-1} D_{22} & D_{21}^{-1} \end{array}\right]$$

and

$$\mathcal{C}_l(M) = \left[\begin{array}{c|ccc} A - B_2 D_{12}^{-1} C_1 & B_2 D_{12}^{-1} & B_1 - B_2 D_{12}^{-1} D_{11} \\ \hline -D_{12}^{-1} C_1 & D_{12}^{-1} & -D_{12}^{-1} D_{11} \\ C_2 - D_{22} D_{12}^{-1} C_1 & D_{22} D_{12}^{-1} & D_{21} - D_{22} D_{12}^{-1} D_{11} \end{array}\right],$$

provided that D_{21} and D_{12} are invertible. When neither D_{21} nor D_{12} is invertible, techniques have to be applied so that the chain-scattering approach can be used [58], but this is irrelevant here.

(viii) Redheffer star product

Consider the system shown in Figure 3.6, which consists of two interconnected subsystems in the input/output representation:

$$\begin{bmatrix} z \\ y \end{bmatrix} = P \begin{bmatrix} w \\ u \end{bmatrix}, \qquad \begin{bmatrix} u \\ \tilde{z} \end{bmatrix} = K \begin{bmatrix} y \\ \tilde{w} \end{bmatrix}.$$

Suppose that P and K are compatibly partitioned matrices

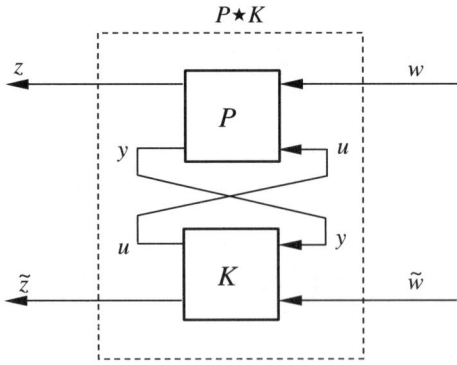

Figure 3.6. Star product of interconnected systems

$$P = \begin{bmatrix} P_{11} & P_{12} \\ P_{21} & P_{22} \end{bmatrix}, \quad K = \begin{bmatrix} K_{11} & K_{12} \\ K_{21} & K_{22} \end{bmatrix}$$

such that $P_{22}K_{11}$ is well defined and square. If $I - P_{22}K_{11}$ is invertible, then the transfer function from $\begin{bmatrix} w \\ \tilde{w} \end{bmatrix}$ to $\begin{bmatrix} z \\ \tilde{z} \end{bmatrix}$ is defined as the *star product* of P and K with respect to this partition, which is given by

$$P \star K = \begin{bmatrix} \mathcal{F}_l(P, K_{11}) & P_{12}(I - K_{11}P_{22})^{-1}K_{12} \\ K_{21}(I - P_{22}K_{11})^{-1}P_{21} & \mathcal{F}_u(K, P_{22}) \end{bmatrix}.$$

Assume that the state-space representations of P and K are

$$P = \left[\begin{array}{c|cc} A & B_1 & B_2 \\ \hline C_1 & D_{11} & D_{12} \\ C_2 & D_{21} & D_{22} \end{array} \right], \quad K = \left[\begin{array}{c|cc} A_K & B_{K1} & B_{K2} \\ \hline C_{K1} & D_{K11} & D_{K12} \\ C_{K2} & D_{K21} & D_{K22} \end{array} \right].$$

Then the state-space representation of the star product is

$$P \star K = \left[\begin{array}{c|cc} \bar{A} & \bar{B}_1 & \bar{B}_2 \\ \hline \bar{C}_1 & \bar{D}_{11} & \bar{D}_{12} \\ \bar{C}_2 & \bar{D}_{21} & \bar{D}_{22} \end{array} \right] = \left[\begin{array}{c|c} \bar{A} & \bar{B} \\ \hline \bar{C} & \bar{D} \end{array} \right],$$

where

$$\bar{A} = \begin{bmatrix} A + B_2 \tilde{R}^{-1} D_{K11} C_2 & B_2 \tilde{R}^{-1} C_{K1} \\ B_{K1} R^{-1} C_2 & A_K + B_{K1} R^{-1} D_{22} C_{K1} \end{bmatrix},$$

$$\bar{B} = \begin{bmatrix} B_1 + B_2 \tilde{R}^{-1} D_{K11} D_{21} & B_2 \tilde{R}^{-1} D_{K12} \\ B_{K1} R^{-1} D_{21} & B_{K2} + B_{K1} R^{-1} D_{22} D_{K12} \end{bmatrix},$$

$$\bar{C} = \begin{bmatrix} C_1 + D_{12} D_{K11} R^{-1} C_2 & D_{12} \tilde{R}^{-1} C_{K1} \\ D_{K21} R^{-1} C_2 & C_{K2} + D_{K21} R^{-1} D_{22} C_{K1} \end{bmatrix},$$

$$\bar{D} = \begin{bmatrix} D_{11} + D_{12} D_{K11} R^{-1} D_{21} & D_{12} \tilde{R}^{-1} D_{K12} \\ D_{K21} R^{-1} D_{21} & D_{K22} + D_{K21} R^{-1} D_{22} D_{K12} \end{bmatrix},$$

with

$$R = I - D_{22} D_{K11}, \quad \tilde{R} = I - D_{K11} D_{22}.$$

(ix) Linear fractional transformations (LFT)

The star product can be used to generate state-space realizations of the upper/lower LFT. Assume that

$$P = \left[\begin{array}{c|cc} A & B_1 & B_2 \\ \hline C_1 & D_{11} & D_{12} \\ C_2 & D_{21} & D_{22} \end{array} \right], \quad K = \left[\begin{array}{c|c} A_K & B_K \\ \hline C_K & D_K \end{array} \right]$$

have appropriate dimensions to carry out the calculations below. Then the state-space realization of $\mathcal{F}_l(P, K)$, if well defined, is

$$\left[\begin{array}{cc|c} A+B_2\tilde{R}^{-1}D_KC_2 & B_2\tilde{R}^{-1}C_K & B_1+B_2\tilde{R}^{-1}D_KD_{21} \\ B_KR^{-1}C_2 & A_K+B_KR^{-1}D_{22}C_K & B_KR^{-1}D_{21} \\ \hline C_1+D_{12}D_KR^{-1}C_2 & D_{12}\tilde{R}^{-1}C_K & D_{11}+D_{12}D_KR^{-1}D_{21} \end{array}\right],$$

where $R = I - D_{22}D_K$ and $\tilde{R} = I - D_KD_{22}$, and that of $\mathcal{F}_u(P, K)$, if well defined, is

$$\left[\begin{array}{cc|cc} A_K+B_K\tilde{R}^{-1}D_{11}C_K & B_K\tilde{R}^{-1}C_1 & B_K\tilde{R}^{-1}D_{12} \\ B_1R^{-1}C_K & A+B_1R^{-1}D_KC_1 & B_2+B_1R^{-1}D_KD_{12} \\ \hline D_{21}R^{-1}C_K & C_2+D_{21}R^{-1}D_KC_1 & D_{22}+D_{21}R^{-1}D_KD_{12} \end{array}\right],$$

where $R = I - D_KD_{11}$ and $\tilde{R} = I - D_{11}D_K$.

If $K = D_K$ is a static feedback gain, then

$$\mathcal{F}_l(P, K) = \left[\begin{array}{c|c} A+B_2D_KR^{-1}C_2 & B_1+B_2D_KR^{-1}D_{21} \\ \hline C_1+D_{12}D_KR^{-1}C_2 & D_{11}+D_{12}D_KR^{-1}D_{21} \end{array}\right]$$

when $R = I - D_{22}D_K$ is invertible, and

$$\mathcal{F}_u(P, K) = \left[\begin{array}{c|c} A+B_1R^{-1}D_KC_1 & B_2+B_1R^{-1}D_KD_{12} \\ \hline C_2+D_{21}R^{-1}D_KC_1 & D_{22}+D_{21}R^{-1}D_KD_{12} \end{array}\right]$$

when $R = I - D_KD_{11}$ is invertible.

(x) Homographic transformations (HMT)

Assume that the state-space representations of N and K are

$$N = \left[\begin{array}{c|cc} A & B_1 & B_2 \\ \hline C_1 & D_{11} & D_{12} \\ C_2 & D_{21} & D_{22} \end{array}\right], \quad K = \left[\begin{array}{c|c} A_K & B_K \\ \hline C_K & D_K \end{array}\right].$$

Then the state-space representation of $\mathcal{H}_r(N, K)$, if well defined, is

$$\mathcal{H}_r(N, K) = \left[\begin{array}{c|c} \bar{A} & \bar{B} \\ \hline \bar{C} & \bar{D} \end{array}\right],$$

where

$$\bar{A} = \begin{bmatrix} A & B_1C_K \\ 0 & A_K \end{bmatrix} - \begin{bmatrix} \hat{B} \\ B_K \end{bmatrix} D_2^{-1} \begin{bmatrix} C_2 & D_{21}C_K \end{bmatrix}$$

$$\bar{B} = \begin{bmatrix} \hat{B} \\ B_K \end{bmatrix} D_2^{-1}$$

$$\bar{C} = \begin{bmatrix} C_1 - D_1D_2^{-1}C_2 & (D_{11} - D_1D_2^{-1}D_{21})C_K \end{bmatrix}$$

$$\bar{D} = D_1D_2^{-1},$$

with

$$\begin{bmatrix} \hat{B} \\ D_1 \\ D_2 \end{bmatrix} = \begin{bmatrix} B_1 & B_2 \\ D_{11} & D_{12} \\ D_{21} & D_{22} \end{bmatrix} \begin{bmatrix} D_K \\ I \end{bmatrix}.$$

3.3.2 Similarity Transformations

In general, the systems resulting after the above operations are not minimal and hence similarity transformations are needed to minimise the state-space realizations, *i.e.*, to remove the unobservable and/or uncontrollable states.

It is well known that, for a nonsingular T,

$$G = \left[\begin{array}{c|c} A & B \\ \hline C & D \end{array}\right] = \left[\begin{array}{c|c} T^{-1}AT & T^{-1}B \\ \hline CT & D \end{array}\right].$$

This is called a *similarity transformation* with T, which corresponds to the change of state variables. It can be done by using elementary column (row) operations followed by the corresponding elementary row (column) operations, as shown in Figure 3.7. The basic operations of similarity transformations are summarised in Table 3.3. If the transformed system has some unobservable and/or uncontrollable states, then they can be removed by deleting the corresponding rows in $\begin{bmatrix} A & B \end{bmatrix}$ and columns in $\begin{bmatrix} A \\ C \end{bmatrix}$ so that the dimension of the realization is reduced. For example, if all the elements on the same row in $\begin{bmatrix} A & B \end{bmatrix}$ or the same column in $\begin{bmatrix} A \\ C \end{bmatrix}$, except the one on the main diagonal of A, are 0, then this row and column can be removed.

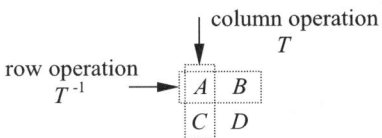

Figure 3.7. Similarity transformation on a system G with T

Table 3.3. Basic similarity transformation operations on a system

elementary row operations		elementary column operations
to multiply some row(s) with a nonsingular matrix (number) T^{-1} from the left	∼	to multiply the corresponding column(s) with a matrix (number) T from the right
to swap rows j and k	∼	to swap columns k and j
to add T times row j (from the left) to row k	∼	to subtract column j with column k times T (from the right)

As can be seen in the following chapters that similarity transformation plays a very important role in this book and forms one of its foundations.

3.4 Algebraic Riccati Equations

3.4.1 Definitions

Let A, R and E be real $n \times n$ matrices[1] with R and E symmetric. The following matrix equation is called an *algebraic Riccati equation* (ARE) [15, 62, 63]:

$$A^*X + XA + XRX + E = 0. \qquad (3.2)$$

It plays an important role in linear system theory, in particular, in H^2 and H^∞ optimal control. The above ARE can also be written in the matrix form

$$\begin{bmatrix} -X & I \end{bmatrix} \begin{bmatrix} A & R \\ -E & -A^* \end{bmatrix} \begin{bmatrix} I \\ X \end{bmatrix} = 0. \qquad (3.3)$$

A matrix of the form $H = \begin{bmatrix} A & R \\ -E & -A^* \end{bmatrix}$ is called a *Hamiltonian matrix*. The spectrum of a Hamiltonian matrix H is symmetric about the imaginary axis because H is similar to $-H^*$:

$$\bar{J}^{-1} H \bar{J} = -\bar{J} H \bar{J} = -H^*,$$

where $\bar{J} = \begin{bmatrix} 0 & -I \\ I & 0 \end{bmatrix}$ having the property $\bar{J}^{-1} = -\bar{J} = \bar{J}^*$. This is often called the *Hamiltonian property* [62].

3.4.2 Stabilising Solution

Since ARE is a quadratic equation, there are many solutions [62]. In control engineering, particular interest is paid to one such that X is real and $A + RX$ is stable. This solution is symmetric [180] and is called the stabilising solution.

Lemma 3.3. *The ARE (3.2) or (3.3) has a stabilising solution X only if H does not have eigenvalues on the $j\omega$-axis.*

Proof. As a matter of fact, Equation (3.3) is part of the following similarity transformation applied to H with $\begin{bmatrix} I & 0 \\ X & I \end{bmatrix}$:

$$\begin{bmatrix} I & 0 \\ -X & I \end{bmatrix} \begin{bmatrix} A & R \\ -E & -A^* \end{bmatrix} \begin{bmatrix} I & 0 \\ X & I \end{bmatrix} = \begin{bmatrix} A + RX & R \\ 0 & -(A+RX)^* \end{bmatrix}, \qquad (3.4)$$

where the $(2,1)$-block is set to 0. If $A + RX$ is stable then $-(A+RX)^*$ is antistable. In other words, H does not have eigenvalues on the $j\omega$-axis. □

However, it is not sufficient for the ARE to have a stabilising solution if H does not have eigenvalues on the $j\omega$-axis. A stronger condition is needed.

---
[1] In Section 3.4, all the block matrices are square, having dimension $n \times n$ or $2n \times 2n$.

3.4 Algebraic Riccati Equations

Lemma 3.4. *Suppose H has no imaginary eigenvalues and R is either positive semi-definite or negative semi-definite. Then a stabilising solution X exists if and only if (A, R) is stabilisable. Furthermore, X is real, symmetric and unique.*

Proof. See [180]. Only the uniqueness is shown here. Let X_1 and X_2 be solutions of (3.2) such that $A + RX_1$ and $A + RX_2$ are stable. Then,
$$A^*X_i + X_iA + X_iRX_i + E = 0 \qquad (i = 1, 2).$$
Subtract one from the other, then
$$(X_1 - X_2)(A + RX_1) + (A + RX_2)^*(X_1 - X_2) = 0.$$
Since $A + RX_1$ and $A + RX_2$ are all stable, there is $X_1 - X_2 = 0$ [58]. □

The first column of (3.4) is
$$\begin{bmatrix} I & 0 \\ -X & I \end{bmatrix} \begin{bmatrix} A & R \\ -E & -A^* \end{bmatrix} \begin{bmatrix} I \\ X \end{bmatrix} = \begin{bmatrix} A + RX \\ 0 \end{bmatrix} = \begin{bmatrix} I \\ 0 \end{bmatrix} (A + RX),$$
where $A + RX$ is stable. This gives
$$\begin{bmatrix} A & R \\ -E & -A^* \end{bmatrix} \begin{bmatrix} I \\ X \end{bmatrix} = \begin{bmatrix} I \\ X \end{bmatrix} (A + RX).$$
In general, a Hamiltonian matrix H is said to belong to $dom(Ric)$ if there exist an X and a stable A_X such that
$$H \begin{bmatrix} I \\ X \end{bmatrix} = \begin{bmatrix} I \\ X \end{bmatrix} A_X. \tag{3.5}$$
This X is the stabilising solution of the corresponding ARE
$$\begin{bmatrix} -X & I \end{bmatrix} H \begin{bmatrix} I \\ X \end{bmatrix} = 0, \tag{3.6}$$
which is obtained by pre-multiplying (3.5) with $\begin{bmatrix} -X & I \end{bmatrix}$, and is denoted by $X = Ric(H)$. The stable A_X can be recovered from (3.5) as
$$A_X = \begin{bmatrix} I & 0 \end{bmatrix} H \begin{bmatrix} I \\ X \end{bmatrix}. \tag{3.7}$$
In this book, if not explicitly specified, a solution of ARE refers to the stabilising solution $X = Ric(H)$. The formulae (3.6) and (3.7) will be represented as a block diagram in the next subsection.

Suppose that H has no imaginary eigenvalues. Then there always exists a nonsingular matrix $T = \begin{bmatrix} X_1 & ? \\ X_2 & ? \end{bmatrix}$, e.g., via the Schur decomposition [63], such that[2]
$$T^{-1}HT = \begin{bmatrix} A_- & ? \\ 0 & ? \end{bmatrix}, \tag{3.8}$$
where A_- contains all the stable eigenvalues of H.

---
[2] The elements denoted by "?" are irrelevant.

Lemma 3.5. *Suppose that H has no imaginary eigenvalues. Then a stabilising solution $X = Ric(H)$ exists if and only if the $(1,1)$-block X_1 of T in (3.8) is nonsingular. Furthermore, the stabilising solution is $X = X_2 X_1^{-1}$.*

Proof. Sufficiency. According to (3.8),

$$H \begin{bmatrix} X_1 & ? \\ X_2 & ? \end{bmatrix} \begin{bmatrix} I \\ 0 \end{bmatrix} = \begin{bmatrix} X_1 & ? \\ X_2 & ? \end{bmatrix} \begin{bmatrix} A_- & ? \\ 0 & ? \end{bmatrix} \begin{bmatrix} I \\ 0 \end{bmatrix},$$

i.e.,

$$H \begin{bmatrix} X_1 \\ X_2 \end{bmatrix} = \begin{bmatrix} X_1 & ? \\ X_2 & ? \end{bmatrix} \begin{bmatrix} A_- \\ 0 \end{bmatrix} = \begin{bmatrix} X_1 & ? \\ X_2 & ? \end{bmatrix} \begin{bmatrix} I \\ 0 \end{bmatrix} A_- = \begin{bmatrix} X_1 \\ X_2 \end{bmatrix} A_-.$$

If X_1 is nonsingular, then

$$H \begin{bmatrix} X_1 \\ X_2 \end{bmatrix} X_1^{-1} = \begin{bmatrix} X_1 \\ X_2 \end{bmatrix} A_- X_1^{-1},$$

which gives

$$H \begin{bmatrix} I \\ X_2 X_1^{-1} \end{bmatrix} = \begin{bmatrix} I \\ X_2 X_1^{-1} \end{bmatrix} X_1 A_- X_1^{-1}.$$

Since $X_1 A_- X_1^{-1} \sim A_-$ and A_- is stable, $H \in dom(Ric)$. The above formulae says that $X = X_2 X_1^{-1} = Ric(H)$ is the stabilising solution.

Necessity. Assume that X is the stabilising solution, then the nonsingular matrix

$$T = \begin{bmatrix} I & 0 \\ X & I \end{bmatrix}$$

satisfies (3.8) because of (3.5). The $(1,1)$-block of T is of course nonsingular. $\square$

Remark 3.1. This lemma reveals the close relationship between ARE and similarity transformation. Once a Hamiltonian matrix H having no imaginary eigenvalues is similarly transformed with T into an upper triangular form with the $(1,1)$-block having all the stable eigenvalues of H, the existence of a stabilising solution to the ARE is equivalent to the nonsingularity of the $(1,1)$-block of the nonsingular matrix T used for the transformation.

3.4.3 Block-diagram Representation

The formulae (3.6) and (3.7) can be represented as the block diagram shown in Figure 3.8, where the matrices U, V, W, Y, U_1, V_1, W_1, Y_1 and X are all square with the same dimension. The block H is the Hamiltonian matrix and the dashed box is called the solution generator. The following relationships hold:

3.4 Algebraic Riccati Equations

$$\begin{bmatrix} W \\ Y \end{bmatrix} = H \begin{bmatrix} U \\ V \end{bmatrix},$$

$$\begin{bmatrix} W_1 \\ Y_1 \end{bmatrix} = \begin{bmatrix} I & 0 \\ -X & I \end{bmatrix} \begin{bmatrix} W \\ Y \end{bmatrix},$$

$$\begin{bmatrix} U \\ V \end{bmatrix} = \begin{bmatrix} I & 0 \\ X & I \end{bmatrix} \begin{bmatrix} U_1 \\ V_1 \end{bmatrix}. \tag{3.9}$$

What is described in this block diagram is actually the similarity transformation done to H with $\begin{bmatrix} I & 0 \\ X & I \end{bmatrix}$. It is assumed that U_1 is nonsingular and V_1 is set to 0. Then,

$$\begin{bmatrix} W_1 \\ Y_1 \end{bmatrix} = \begin{bmatrix} I & 0 \\ -X & I \end{bmatrix} H \begin{bmatrix} I & 0 \\ X & I \end{bmatrix} \begin{bmatrix} U_1 \\ 0 \end{bmatrix} = \begin{bmatrix} I & 0 \\ -X & I \end{bmatrix} H \begin{bmatrix} I \\ X \end{bmatrix} U_1.$$

Hence, the transfer function from U_1 to W_1 is the A_X given in (3.7) and that from U_1 to Y_1 is the ARE given in (3.6), which is equal to 0.

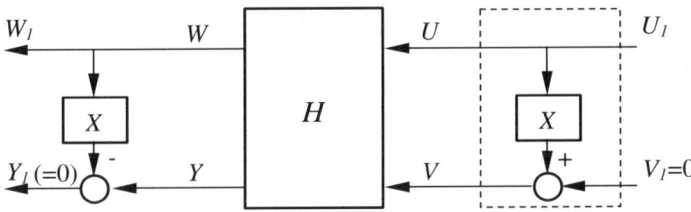

Figure 3.8. Block-diagram representation of algebraic Riccati equation

The solution to the ARE can also be generated from the block diagram. Substituting $V_1 = 0$ into (3.9), the dashed block (called the solution generator) can be described as follows

$$U = U_1,$$
$$V = XU_1 = XU.$$

The solution X is the transfer gain from U to V, i.e.,

$$X = VU^{-1},$$

assuming U is invertible.

3.4.4 Similarity Transformations and Stabilising Solutions

In general, a similarity transform does not preserve the Hamiltonian property. However, some similarity transforms do, in particular, when the trans-

(i) Parallel structure

Lemma 3.6. *Let H_0 be a Hamiltonian matrix with $X_0 = Ric(H_0)$. Then $H = T^{-1}H_0T$ is still Hamiltonian for any $T = \begin{bmatrix} I & 0 \\ L & I \end{bmatrix}$ with $L = L^*$. Furthermore, $X = Ric(H)$ exists and is given by*

$$X = X_0 - L.$$

Proof. Note that

$$T^{-1}\bar{J}T^{-*} = \begin{bmatrix} I & 0 \\ -L & I \end{bmatrix} \begin{bmatrix} 0 & -I \\ I & 0 \end{bmatrix} \begin{bmatrix} I & -L \\ 0 & I \end{bmatrix} = \begin{bmatrix} 0 & -I \\ I & 0 \end{bmatrix} = \bar{J}. \tag{3.10}$$

Substitute $H_0 = THT^{-1}$ into $H_0\bar{J} = -\bar{J}H_0^*$, then

$$THT^{-1}\bar{J} = -\bar{J}(THT^{-1})^* = -\bar{J}T^{-*}H^*T^*,$$

which gives

$$HT^{-1}\bar{J}T^{-*} = -T^{-1}\bar{J}T^{-*}H^*.$$

According to (3.10),

$$H\bar{J} = -\bar{J}H^* = (H\bar{J})^*.$$

This means H is Hamiltonian.

Since $X_0 = Ric(H_0)$, according to (3.5),

$$H_0 \begin{bmatrix} I \\ X_0 \end{bmatrix} = \begin{bmatrix} I \\ X_0 \end{bmatrix} A_{X0},$$

where A_{X0} is stable. Then,

$$THT^{-1} \begin{bmatrix} I \\ X_0 \end{bmatrix} = \begin{bmatrix} I \\ X_0 \end{bmatrix} A_{X0}.$$

This gives

$$H \begin{bmatrix} I \\ X \end{bmatrix} = \begin{bmatrix} I \\ X \end{bmatrix} A_{X0}, \qquad X = X_0 - L.$$

Hence, $X = Ric(H) = X_0 - L$. □

---

[3] If λ is an eigenvalue of a symplectic matrix, then λ^{-1} is as well. A matrix $\begin{bmatrix} A & B \\ C & D \end{bmatrix}$ with square block-matrices A, B, C and D is symplectic if and only if $AD^T - BC^T = I$, $AB^T = BA^T$ and $CD^T = DC^T$.

3.4 Algebraic Riccati Equations

This can be described as the block diagram shown in Figure 3.9. The left side from the Hamiltonian matrix H_0 is omitted here. The two L blocks cancel each other because $TT^{-1} = I$. It is clear from Figure 3.8 that Figure 3.9 describes $X_0 = Ric(H_0)$. However, if the two L blocks are split, then the left side gives $H = \begin{bmatrix} I & 0 \\ -L & I \end{bmatrix} H_0 \begin{bmatrix} I & 0 \\ L & I \end{bmatrix}$ and the right side, the dashed block, generates the solution

$$X = Ric(H) = X_0 - L$$

because

$$U = U_1, \qquad V = (X_0 - L)U_1 = (X_0 - L)U.$$

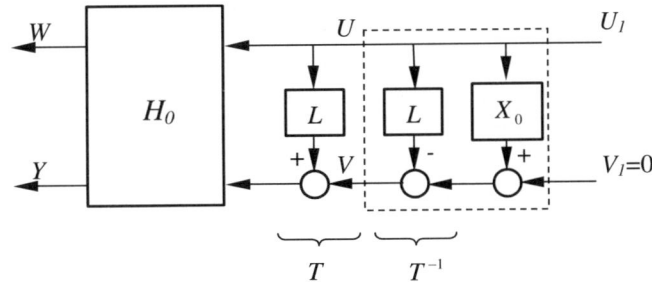

Figure 3.9. Solution generator when $T = \begin{bmatrix} I & 0 \\ L & I \end{bmatrix}$

(ii) Series structure

Lemma 3.7. *Let H_0 be a Hamiltonian matrix with $X_0 = Ric(H_0)$. Then $H = T^{-1}H_0T$ is still Hamiltonian for $T = \begin{bmatrix} L & 0 \\ 0 & L^{-T} \end{bmatrix}$ with a nonsingular L. Furthermore, $X = Ric(H)$ exists and is given by*

$$X = L^T X_0 L.$$

Proof. Repeat the proof of Lemma 3.6 with $T = \begin{bmatrix} L & 0 \\ 0 & L^{-T} \end{bmatrix}$, noting that

$$T^{-1}\bar{J}T^{-*} = \begin{bmatrix} L^{-1} & 0 \\ 0 & L^T \end{bmatrix} \begin{bmatrix} 0 & -I \\ I & 0 \end{bmatrix} \begin{bmatrix} L^{-T} & 0 \\ 0 & L \end{bmatrix} = \begin{bmatrix} 0 & -I \\ I & 0 \end{bmatrix} = \bar{J}$$

and that

$$H \begin{bmatrix} I \\ L^T X_0 L \end{bmatrix} = \begin{bmatrix} I \\ L^T X_0 L \end{bmatrix} L^{-1} A_{X_0} L.$$

□

64   3 Preliminaries

In this case, the solution generator is shown in Figure 3.10. It consists of a block T^{-1} and the original solution block X_0. Since

$$U = L^{-1}U_1, \quad V = L^T X_0 U_1,$$

the solution $X = Ric(H) = Ric(T^{-1}H_0 T)$ is

$$X = VU^{-1} = L^T X_0 L.$$

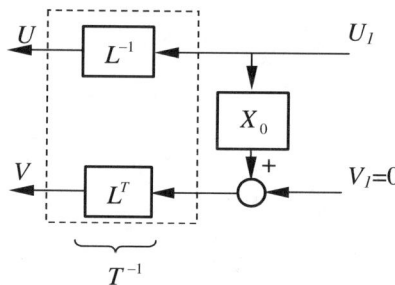

Figure 3.10. Solution generator when $T = \begin{bmatrix} L & 0 \\ 0 & L^{-T} \end{bmatrix}$

(iii) Feedback structure

Lemma 3.8. *Let H_0 be a Hamiltonian matrix with $X_0 = Ric(H_0)$. Then $H = T^{-1}H_0 T$ is still Hamiltonian for $T = \begin{bmatrix} I & L \\ 0 & I \end{bmatrix}$, $L = L^*$. Furthermore, $X = Ric(H)$ exists when $I - LX_0$ is nonsingular and is given by*

$$X = Ric(H) = X_0(I - LX_0)^{-1} = (I - X_0 L)^{-1} X_0.$$

Proof. Repeat the proof of Lemma 3.6 with $T = \begin{bmatrix} I & L \\ 0 & I \end{bmatrix}$, noting that

$$T^{-1}\bar{J}T^{-*} = \begin{bmatrix} I & -L \\ 0 & I \end{bmatrix} \begin{bmatrix} 0 & -I \\ I & 0 \end{bmatrix} \begin{bmatrix} I & 0 \\ -L & I \end{bmatrix} = \begin{bmatrix} 0 & -I \\ I & 0 \end{bmatrix} = \bar{J}$$

and that

$$H \begin{bmatrix} I - LX_0 \\ X_0 \end{bmatrix} = \begin{bmatrix} I - LX_0 \\ X_0 \end{bmatrix} A_{X0}.$$

When $I - LX_0$ is nonsingular, the latter is equivalent to

$$H \begin{bmatrix} I \\ X_0(I - LX_0)^{-1} \end{bmatrix} = \begin{bmatrix} I \\ X_0(I - LX_0)^{-1} \end{bmatrix} (I - LX_0) A_{X0} (I - LX_0)^{-1}.$$

□

3.4 Algebraic Riccati Equations

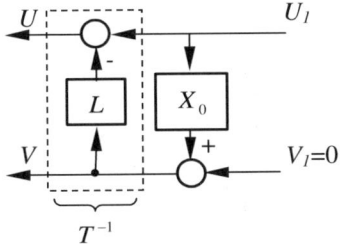

Figure 3.11. Solution generator when $T = \begin{bmatrix} I & L \\ 0 & I \end{bmatrix}$

In this case, the solution generator is shown in Figure 3.11. It consists of a block T^{-1} and the original solution block X_0. Since

$$U = (I - LX_0)U_1, \qquad V = X_0 U_1,$$

the solution $X = Ric(H) = Ric(T^{-1}H_0T)$ is

$$X = VU^{-1} = X_0(I - LX_0)^{-1}$$

when $I - LX_0$ is nonsingular.

(iv) Inverse structure

Lemma 3.9. *Let H_0 be a Hamiltonian matrix with $X_0 = Ric(H_0)$. Then $H = T^{-1}H_0T$ is still Hamiltonian for $T = \begin{bmatrix} 0 & I \\ I & 0 \end{bmatrix}$. Furthermore, $X = Ric(H)$ exists when X_0 is invertible and is given by*

$$X = X_0^{-1}.$$

Proof. Repeat the proof of Lemma 3.6 with $T = \begin{bmatrix} 0 & I \\ I & 0 \end{bmatrix}$, noting that

$$T^{-1}\bar{J}T^{-*} = \begin{bmatrix} 0 & I \\ I & 0 \end{bmatrix} \begin{bmatrix} 0 & -I \\ I & 0 \end{bmatrix} \begin{bmatrix} 0 & I \\ I & 0 \end{bmatrix} = \begin{bmatrix} 0 & I \\ -I & 0 \end{bmatrix} = -\bar{J}$$

and that

$$H \begin{bmatrix} X_0 \\ I \end{bmatrix} = \begin{bmatrix} X_0 \\ I \end{bmatrix} A_{X0}.$$

When X_0 is invertible, the latter is equivalent to

$$H \begin{bmatrix} I \\ X_0^{-1} \end{bmatrix} = \begin{bmatrix} I \\ X_0^{-1} \end{bmatrix} X_0 A_{X0} X_0^{-1}.$$

□

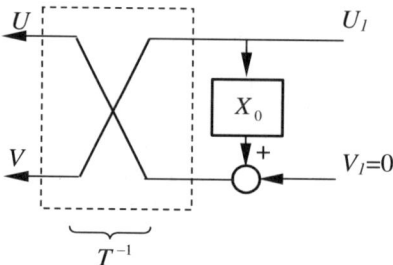

Figure 3.12. Solution generator when $T = \begin{bmatrix} 0 & I \\ I & 0 \end{bmatrix}$

In this case, the solution generator is shown in Figure 3.12. Since

$$U = X_0 U_1, \qquad V = U_1,$$

the solution $X = Ric(H) = Ric(T^{-1} H_0 T)$ is

$$X = VU^{-1} = X_0^{-1}$$

when X_0 is nonsingular.

Remark 3.2. Roughly speaking, after swapping rows and columns of a Hamiltonian matrix, the stabilising solution of the corresponding ARE becomes the inverse of that of the original ARE, if invertible.

(v) Opposite structure

Lemma 3.10. *Let H_0 be a Hamiltonian matrix with $X_0 = Ric(H_0)$. Then $H = T^{-1} H_0 T$ is still Hamiltonian for $T = \begin{bmatrix} -I & 0 \\ 0 & I \end{bmatrix}$ or $T = \begin{bmatrix} I & 0 \\ 0 & -I \end{bmatrix}$. Furthermore, $X = Ric(H)$ exists and is given by*

$$X = -X_0.$$

Proof. Repeat the proof of Lemma 3.6 with $T = \begin{bmatrix} -I & 0 \\ 0 & I \end{bmatrix}$ or $T = \begin{bmatrix} I & 0 \\ 0 & -I \end{bmatrix}$, noting that

$$T^{-1} \bar{J} T^{-*} = \begin{bmatrix} 0 & I \\ -I & 0 \end{bmatrix} = -\bar{J}$$

and that

$$H \begin{bmatrix} I \\ -X_0 \end{bmatrix} = \begin{bmatrix} I \\ -X_0 \end{bmatrix} A_{X0}.$$

□

3.4 Algebraic Riccati Equations

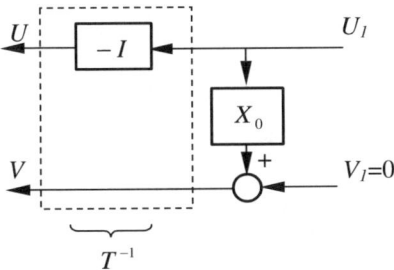

Figure 3.13. Solution generator when $T = \begin{bmatrix} -I & 0 \\ 0 & I \end{bmatrix}$

When $T = \begin{bmatrix} -I & 0 \\ 0 & I \end{bmatrix}$, the solution generator is shown in Figure 3.13. It is similar when $T = \begin{bmatrix} I & 0 \\ 0 & -I \end{bmatrix}$. Since

$$U = -U_1, \qquad V = X_0 U_1,$$

the solution $X = Ric(H) = Ric(T^{-1}H_0T)$ is

$$X = VU^{-1} = -X_0.$$

Remark 3.3. This lemma says that if both R and E in an ARE change sign then the solution of the ARE changes sign as well.

For combined similarity transforms, the stabilising solution of the ARE can be constructed according to the above structures.

3.4.5 Rank Defect of Stabilising Solutions

Not all stabilising solutions are of full rank. This is related to the structural property of H.

Lemma 3.11. *Assume* $H = \begin{bmatrix} A & R \\ -E & -A^* \end{bmatrix}$ *is a Hamiltonian matrix with* $E \leq 0$ *and* $X = Ric(H)$ *is singular. Then any nonsingular matrix* L *that satisfies*

$$L^T X L = \begin{bmatrix} \bar{X} & 0 \\ 0 & 0 \end{bmatrix}, \qquad \bar{X} \text{ is nonsingular}$$

transforms each matrix in H *into the following form with appropriate dimensions compatible with* $\bar{X}$:

$$L^{-1}AL = \begin{bmatrix} A_{11} & 0 \\ A_{21} & A_{22} \end{bmatrix}, \qquad A_{22} \text{ is stable}$$

$$-L^T EL = \begin{bmatrix} -\bar{E} & 0 \\ 0 & 0 \end{bmatrix},$$

$$L^{-1}RL^{-T} = \begin{bmatrix} R_{11} & R_{12} \\ R_{12}^* & R_{22} \end{bmatrix}.$$

Also, $\bar{X} = Ric(\bar{H})$ with

$$\bar{H} = \begin{bmatrix} A_{11} & R_{11} \\ -\bar{E} & -A_{11}^* \end{bmatrix}.$$

Proof. Apply Lemma 3.7. See [58, Lemma 3.5] for details. □

3.4.6 Stabilising or Grouping?

A solution X of the ARE (3.2) is often called stabilising if $A + RX$ is stable. It does not mean to stabilise A via X because A itself may be stable. Even if A is unstable, it does not mean to simply stabilise it via X. A better term is *grouping*: X is a grouping solution such that all the stable eigenvalues of H are grouped into $A + RX$ and all the unstable eigenvalues of H are grouped into $-(A + RX)^*$. See (3.4) or the lemma below. Once the eigenvalues are grouped, they can be factorised. This is why the factorisation of a transfer function, which is very common in advanced control theory, always boils down to solving some algebraic Riccati equations.

Lemma 3.12. *Let $H = \begin{bmatrix} A & R \\ -E & -A^* \end{bmatrix}$ be a Hamiltonian matrix with $X = Ric(H)$ and P be the solution of the Lyapunov equation*

$$A_X P + P A_X^* + R = 0,$$

where $A_X = A + RX$ is stable. Then the nonsingular matrix

$$T = \begin{bmatrix} I & 0 \\ X & I \end{bmatrix} \begin{bmatrix} I & P \\ 0 & I \end{bmatrix} = \begin{bmatrix} I & P \\ X & I + XP \end{bmatrix}$$

similarly transforms H into the block diagonal form given below:

$$T^{-1}HT = \begin{bmatrix} A_X & 0 \\ 0 & -A_X^* \end{bmatrix}.$$

Proof. It is straightforward to show that the two similarity transformations with $\begin{bmatrix} I & 0 \\ X & I \end{bmatrix}$ and $\begin{bmatrix} I & P \\ 0 & I \end{bmatrix}$ bring H into the diagonal form. □

Consider the ARE (3.3) with $A = -1$, $R = -1$ and $E = 3$. The eigenvalues of the corresponding Hamiltonian matrix $H = \begin{bmatrix} -1 & -1 \\ -3 & 1 \end{bmatrix}$ are $\lambda_{1,2} = \pm 2$. The ARE has two solutions: $X_1 = 1$ and $X_2 = -3$. The stabilising solution is $X_1 = 1$ because $A + RX_1 = -2$ is stable but $A + RX_2 = 2$ is not. Apparently, there is no need to stabilise A here. The function of the solution $X = X_1 = 1$ is to group the stable eigenvalue $\lambda = -2$ of H into $A + RX$: $A + RX = -2$.

3.5 The Σ Matrix

In this section, a very important matrix Σ, which will be used in several subsequent chapters, is discussed.

3.5.1 Definition of the Σ Matrix

Consider the following linear Hamiltonian matrix system:

$$\begin{bmatrix} \dot{U}(t) \\ \dot{V}(t) \end{bmatrix} = \begin{bmatrix} A & \gamma^{-2}BB^* \\ -C^*C & -A^* \end{bmatrix} \begin{bmatrix} U(t) \\ V(t) \end{bmatrix} \quad (3.11)$$

where the system matrix

$$H = \begin{bmatrix} A & \gamma^{-2}BB^* \\ -C^*C & -A^* \end{bmatrix} \quad (3.12)$$

is the A-matrix of $\left(\gamma^2 I - G_{\beta 11}^{\sim} G_{\beta 11}\right)^{-1}$ with[4] $G_{\beta 11} = \left[\begin{array}{c|c} A & B \\ \hline -C & 0 \end{array}\right]$. The transition matrix of the Hamiltonian system (3.11) from $t = 0$ to an arbitrary time t is

$$\Sigma(t) = \exp\left(\begin{bmatrix} A & \gamma^{-2}BB^* \\ -C^*C & -A^* \end{bmatrix} t\right),$$

and the solution to the initial state $\begin{bmatrix} U(0) \\ V(0) \end{bmatrix}$ is

$$\begin{bmatrix} U(t) \\ V(t) \end{bmatrix} = \Sigma(t) \begin{bmatrix} U(0) \\ V(0) \end{bmatrix}. \quad (3.13)$$

In this book, the transition matrix from $t = 0$ to $t = h$,

$$\Sigma = \begin{bmatrix} \Sigma_{11} & \Sigma_{12} \\ \Sigma_{21} & \Sigma_{22} \end{bmatrix} \doteq \Sigma(h) = e^{Hh}, \quad (3.14)$$

is called the Σ *matrix*. As will be shown later, this exponential Hamiltonian matrix plays quite an important role in the H^∞-control of dead-time systems. It is the A-matrix of the modified Smith predictor and it is also closely related to the $L_2[0, h]$-induced norm of $G_{\beta 11}$, as will be shown in the next subsection, and to the solvability condition of several problems for systems with a single delay.

The system (3.11) is intimately connected to the differential Riccati equation[5]

$$\dot{X}(t) = \begin{bmatrix} I & -X(t) \end{bmatrix} H \begin{bmatrix} X(t) \\ I \end{bmatrix}, \quad (3.15)$$

or equivalently,

$$\dot{X}(t) = AX(t) + X(t)A^* + \gamma^{-2}BB^* + X(t)C^*CX(t).$$

---
[4] The reason for use of the notation $G_{\beta 11}$ is to maintain consistency with later chapters.
[5] Note that the right side of (3.15) is in the dual form of the conventional Riccati equation.

The solution of (3.15) can be given using the following well-known result [15, Chapter 3]:

If U, V is a solution pair of (3.11) with V nonsingular on a t-interval T, then $X = UV^{-1}$ is a solution of (3.15) on T. Conversely, if X is a solution of (3.15) on T and V is a fundamental solution of $\dot{V} = -(A^* + C^*CX)V$, then $U = XV, V$ is a solution pair of (3.11) on T. Hence, the following equality holds if $[0, h] \subset T$ and V is nonsingular:

$$X(h) = (\Sigma_{11}X(0) + \Sigma_{12})(\Sigma_{21}X(0) + \Sigma_{22})^{-1} = \mathcal{H}_r(\Sigma, X(0)), \quad (3.16)$$

where $X(0) = U(0)V^{-1}(0)$, or, equivalently,

$$X(0) = (\Sigma_{22}^*X(h) - \Sigma_{12}^*)(-\Sigma_{21}^*X(h) + \Sigma_{11}^*)^{-1} = \mathcal{H}_r(\Sigma^{-1}, X(h)).$$

3.5.2 Important Properties of Σ

Lemma 3.13. Σ *is symplectic, i.e., for* $\bar{J} = \begin{bmatrix} 0 & -I \\ I & 0 \end{bmatrix}$

$$\Sigma^{-1} = \bar{J}^{-1}\Sigma^*\bar{J} = \begin{bmatrix} \Sigma_{22}^* & -\Sigma_{12}^* \\ -\Sigma_{21}^* & \Sigma_{11}^* \end{bmatrix}.$$

Proof. Since H given in (3.12) is Hamiltonian,

$$(H\bar{J})^* = -\bar{J}H^* = H\bar{J}.$$

According to (3.14),

$$\Sigma\bar{J} = e^{Hh}\bar{J}$$
$$= \bar{J} + H\bar{J}h + \frac{1}{2!}(Hh)^2\bar{J} + \frac{1}{3!}(Hh)^3\bar{J} + \cdots$$
$$= \bar{J} - \bar{J}H^*h - \frac{1}{2!}H\bar{J}H^*h^2 - \frac{1}{3!}H^2\bar{J}H^*h^3 - \cdots$$
$$= \bar{J} - \bar{J}H^*h + \frac{1}{2!}\bar{J}(H^*h)^2 + \frac{1}{3!}H\bar{J}H^*H^*h^3 + \cdots$$
$$= \bar{J} - \bar{J}H^*h + \frac{1}{2!}\bar{J}(H^*h)^2 - \frac{1}{3!}\bar{J}(H^*h)^3 - \cdots$$
$$= \bar{J}e^{-H^*h} = \bar{J}\Sigma^{-*}.$$

This gives $\Sigma^{-1} = \bar{J}^{-1}\Sigma^*\bar{J}$ or $\Sigma\bar{J}\Sigma^* = \bar{J}$ and, hence, Σ is symplectic. □

Note also that

$$\Sigma = e^{Hh} = e^{\bar{J}^{-1}\bar{J}Hh} = e^{\bar{J}^{-1}H^*\bar{J}^*h} = e^{-\bar{J}^{-1}H^*h\bar{J}}$$
$$= \bar{J}^{-1}e^{-H^*h}\bar{J} = \bar{J}^{-1}\Sigma^{-*}\bar{J} = \bar{J}\Sigma^{-*}\bar{J}^{-1}.$$

3.5 The Σ Matrix

Lemma 3.14. *Some important properties of Σ are summarised below.*
  (i) The following identities hold:

$$\Sigma_{11}\Sigma_{12}^* = \Sigma_{12}\Sigma_{11}^*, \quad \Sigma_{21}\Sigma_{22}^* = \Sigma_{22}\Sigma_{21}^*,$$

$$\Sigma_{22}^*\Sigma_{12} = \Sigma_{12}^*\Sigma_{22} \quad \text{and} \quad \Sigma_{21}^*\Sigma_{11} = \Sigma_{11}^*\Sigma_{21}.$$

In other words, $\Sigma_{11}\Sigma_{12}^*$, $\Sigma_{12}^*\Sigma_{22}$, $\Sigma_{22}\Sigma_{21}^*$, $\Sigma_{21}^*\Sigma_{11}$ are self-adjoint and so are $\Sigma_{11}^{-1}\Sigma_{12}$, $\Sigma_{22}^{-1}\Sigma_{21}$, $\Sigma_{21}\Sigma_{11}^{-1}$ and $\Sigma_{12}\Sigma_{22}^{-1}$ when Σ_{11} and/or Σ_{22} are invertible.
  (ii) $\Sigma \begin{bmatrix} A & 0 \\ -C^*C & -A^* \end{bmatrix} \Sigma^{-1}$ is a Hamiltonian matrix as well.
  (iii) Σ and $(sI - H)^{-1}$ commute with each other, i.e., $\Sigma^{-1}(sI - H)^{-1}\Sigma = (sI - H)^{-1}$.
  (iv) Σ and H commute with each other, i.e., $H = \Sigma H \Sigma^{-1}$.
  (v) The following identities hold (when Σ_{11} or Σ_{22} is nonsingular, if needed):

$$\Sigma_{22}^{-*} = \Sigma_{11} - \Sigma_{12}\Sigma_{22}^{-1}\Sigma_{21},$$

$$\Sigma_{11}^{-*} = \Sigma_{22} - \Sigma_{21}\Sigma_{11}^{-1}\Sigma_{12},$$

$$\Sigma_{22}^*A - A\Sigma_{22}^* + \gamma^{-2}BB^*\Sigma_{21}^* + \Sigma_{12}^*C^*C = 0,$$

$$\Sigma_{22}^*\gamma^{-2}BB^* - \gamma^{-2}BB^*\Sigma_{11}^* + A\Sigma_{12}^* + \Sigma_{12}^*A^* = 0,$$

$$\Sigma_{21}^*A + \Sigma_{11}^*C^*C - C^*C\Sigma_{22}^* + A^*\Sigma_{21}^* = 0,$$

$$\Sigma_{12}^*C^*C\Sigma_{12} + \gamma^{-2}\Sigma_{22}^*BB^*\Sigma_{22} + \Sigma_{12}^*A^*\Sigma_{22} + \Sigma_{22}^*A\Sigma_{12} = \gamma^{-2}BB^*,$$

$$\Sigma_{21}^*A\Sigma_{11} + \Sigma_{11}^*C^*C\Sigma_{11} + \gamma^{-2}\Sigma_{21}^*BB^*\Sigma_{21} + \Sigma_{11}^*A^*\Sigma_{21} = C^*C,$$

$$\Sigma_{22}^*A\Sigma_{11} + \Sigma_{12}^*C^*C\Sigma_{11} + \gamma^{-2}\Sigma_{22}^*BB^*\Sigma_{21} + \Sigma_{12}^*A^*\Sigma_{21} = A.$$

Proof. These properties are mostly due to the fact that Σ is symplectic. They can be proved using straightforward matrix manipulations. Only the last six identities are shown below.

The last three identities can be derived from the matrix identity

$$\begin{bmatrix} \Sigma_{22}^* & -\Sigma_{12}^* \\ -\Sigma_{21}^* & \Sigma_{11}^* \end{bmatrix} \begin{bmatrix} A & \gamma^{-2}BB^* \\ -C^*C & -A^* \end{bmatrix} \begin{bmatrix} \Sigma_{11} & \Sigma_{12} \\ \Sigma_{21} & \Sigma_{22} \end{bmatrix} = \begin{bmatrix} A & \gamma^{-2}BB^* \\ -C^*C & -A^* \end{bmatrix}$$

and the other three identities can be derived from the matrix identity

$$\begin{bmatrix} \Sigma_{22}^* & -\Sigma_{12}^* \\ -\Sigma_{21}^* & \Sigma_{11}^* \end{bmatrix} \begin{bmatrix} A & \gamma^{-2}BB^* \\ -C^*C & -A^* \end{bmatrix} = \begin{bmatrix} A & \gamma^{-2}BB^* \\ -C^*C & -A^* \end{bmatrix} \begin{bmatrix} \Sigma_{22}^* & -\Sigma_{12}^* \\ -\Sigma_{21}^* & \Sigma_{11}^* \end{bmatrix}.$$

□

3.6 The $L_2[0, h]$-induced Norm

The $L_2[0, h]$-induced norm of $G_{\beta 11} = \left[\begin{array}{c|c} A & B \\ \hline -C & 0 \end{array}\right]$ is defined as

$$\gamma_h = \|G_{\beta 11}\|_{L_2[0,h]} = \sup_{\|u\|_{[0,h]} \neq 0} \frac{\|y\|_{[0,h]}}{\|u\|_{[0,h]}},$$

where $y = G_{\beta 11} u$ and

$$\|u\|_{[0,h]} = \sqrt{\int_0^h u^T u \, dt}, \qquad \|y\|_{[0,h]} = \sqrt{\int_0^h y^T y \, dt}.$$

Various methods [30, 41, 182] were proposed to compute this norm. A simple representation of γ_h is the following Zhou–Khargonekar formula [55] as given in [182]:

$$\gamma_h = \max\{\gamma : \det \Sigma_{22} = 0\}$$

i.e., the maximal γ that makes Σ_{22} singular, where Σ_{22} is the (2, 2)-block of the Σ matrix defined in the previous subsection.

Some results related to this norm [41, 182] are summarised below with proofs omitted.

Lemma 3.15. *The following conditions are equivalent:*

(i) $\gamma > \gamma_h$;

(ii) there exists $Q_0 \in H^\infty$ such that $\|T_h(G_{\beta 11}) + e^{-sh} Q_0\|_\infty < \gamma$;

(iii) Σ_{22} is nonsingular not only for γ but also for any number larger than γ;

(iv) there exists a unique function $X(t)$ satisfying the differential Riccati equation (3.15) with $X(0) = 0$ for $t \in [0, h]$. In particular, $X(h) = \Sigma_{12} \Sigma_{22}^{-1}$;

(v) there exists a unique solution pair $U(t), V(t)$ satisfying the differential Riccati equation (3.11) with $U(0) = 0, V(0) = I$ for $t \in [0, h]$. In particular, $U(h) = \Sigma_{12}, V(h) = \Sigma_{22}$.

4

J-spectral Factorisation of Regular Para-Hermitian Transfer Matrices

This chapter[1] characterises a class of regular para-Hermitian transfer matrices and then studies the J-spectral factorisation of this class using similarity transformations. A transfer matrix Λ in this class admits a J-spectral factorisation if and only if there exists a common nonsingular matrix to similarly transform the A-matrices of Λ and Λ^{-1}, resp., into 2×2 lower (upper, resp.) triangular block matrices with the $(1,1)$-block including all the stable modes of Λ (Λ^{-1}, resp.). For a transfer matrix in a smaller subset, this nonsingular matrix is formulated in terms of the stabilising solutions of two algebraic Riccati equations. The J-spectral factor is formulated in terms of the original realization of the transfer matrix. The approach developed here lays one of the foundations for this book and will be used in the next chapter to solve the delay-type Nehari problem.

4.1 Introduction

J-spectral factorisation plays an important role in H_∞ control of finite-dimensional systems [32, 38, 39, 40] as well as infinite-dimensional systems [51, 80, 110]. The necessary and sufficient condition of the J-spectral factorisation has been well understood [12, 32, 39, 74, 119]. The J-spectral factorisations involved in the literature are done for matrices in the form $G^\sim JG$, mostly with a stable G. For the case with an unstable G, the following three steps can be used to find the J-spectral factor of $G^\sim JG$, by applying the results in [74, Corollary 3.1]:
   (i) to find the modal factorisation of $\Lambda = G^\sim JG$;
   (ii) to construct a stable G_- such that $\Lambda = G_-^\sim JG_-$;
   (iii) to derive the J-spectral factor of $G_-^\sim JG_-$, i.e., of $G^\sim JG$.

---

[1] Portions reprinted from [169, 170], with permission from Elsevier.

For example, if $\Lambda = G^\sim JG$ (with G unstable) is factorised as $\Lambda = T + T^\sim$ with T stable, then $G_- = \begin{bmatrix} I + \frac{T}{2} \\ I - \frac{T}{2} \end{bmatrix}$ is stable[2] and $\Lambda = G_-^\sim JG_-$. It can then be factorised by applying Theorem 2.4 in [74]. However, in some cases, a para-Hermitian transfer matrix Λ is given in the form of a state-space realization and cannot be explicitly written in the form $G^\sim JG$, e.g., in the context of H_∞ control of time-delay systems [78, 160]. In order to use the above-mentioned results, one would have to find a G such that $\Lambda = G^\sim JG$. It would be advantageous if this step could be avoided.

A recent work [119] dealt with this problem. A two-step procedure was proposed to find the J-spectral factor in [119]: (i) to transform Λ into an ordered Schur form; and then (ii) to solve an algebraic Riccati equation (ARE) when there is a stabilising solution. There is no need to find a stable G such that $\Lambda = G^\sim JG$ any more. The advantage of this result is that the realization of Λ need not be minimal or in the Hamiltonian structure (because of the first step). This chapter proposes a different approach to deal with the problem. It involves only very elementary mathematical tools, such as similarity transformations, so that it is easy to understand. The approach developed is crucial to solve the delay-type Nehari problem discussed in the next chapter [160, 163].

A better literature review about this topic can be found in [39, 119] and the references therein. For a wider topic, the symmetric factorisation, see [120] and the references therein.

4.2 Properties of Projections

Before discussing the J-spectral factorisation, the projection matrix of a nonorthogonal projection is derived. This is crucial to derive the main results in this chapter.

For a given nonsingular matrix partitioned as $\begin{bmatrix} M & N \end{bmatrix}$, denote the projection onto the subspace $\operatorname{Im} M$ along the subspace $\operatorname{Im} N$ by P. Then,

$$PM = M, \qquad PN = 0,$$

i.e.,

$$P \begin{bmatrix} M & N \end{bmatrix} = \begin{bmatrix} M & 0 \end{bmatrix}.$$

Hence, the projection matrix P is

$$P = \begin{bmatrix} M & 0 \end{bmatrix} \begin{bmatrix} M & N \end{bmatrix}^{-1}.$$

Similarly, the projection Q onto the subspace $\operatorname{Im} N$ along the subspace $\operatorname{Im} M$ is given by

---

[2] For $\Lambda = T + T^\sim$ with a stable T, a stable G_- such that $\Lambda = G_-^\sim JG_-$ is as given above but not $G_- = \frac{1}{2} \begin{bmatrix} I + T \\ I - T \end{bmatrix}$. If the latter is the case, then Λ should be factorised as $\Lambda = \frac{1}{2}(T + T^\sim)$.

$$Q = \begin{bmatrix} 0 & N \end{bmatrix} \begin{bmatrix} M & N \end{bmatrix}^{-1} = \begin{bmatrix} N & 0 \end{bmatrix} \begin{bmatrix} N & M \end{bmatrix}^{-1}.$$

A projection holds a lot of interesting properties; see the excellent book [14]. In particular, the following properties hold for P and Q:
(i) $P + Q = I$;
(ii) $PQ = 0$;
(iii) P and Q are idempotent matrices[3], i.e., $P^2 = P$ and $Q^2 = Q$;
(iv) $\operatorname{Im} P = \operatorname{Im} M$;
(v) $\begin{bmatrix} M & 0 \end{bmatrix} \begin{bmatrix} M & N \end{bmatrix}^{-1} \begin{bmatrix} M & 0 \end{bmatrix} = \begin{bmatrix} M & 0 \end{bmatrix}$;
(vi) $\begin{bmatrix} 0 & N \end{bmatrix} \begin{bmatrix} M & N \end{bmatrix}^{-1} \begin{bmatrix} 0 & N \end{bmatrix} = \begin{bmatrix} 0 & N \end{bmatrix}$;
(vii) $\begin{bmatrix} M & 0 \end{bmatrix} \begin{bmatrix} M & N \end{bmatrix}^{-1} \begin{bmatrix} 0 & N \end{bmatrix} = 0$.

If $M^T N = 0$, i.e., the projection is orthogonal, then

$$\begin{bmatrix} M & N \end{bmatrix}^{-1} = \begin{bmatrix} (M^T M)^{-1} M^T \\ (N^T N)^{-1} N^T \end{bmatrix}.$$

The projection matrices P and Q reduce to

$$P = M(M^T M)^{-1} M^T$$

and

$$Q = N(N^T N)^{-1} N^T.$$

These two formulae can be easily found in the literature.

4.3 Regular Para-Hermitian Transfer Matrices

Definition 4.1. *[60] A transfer matrix $\Lambda(s)$ is called a* para-Hermitian *matrix if $\Lambda^\sim(s) = \Lambda(s)$.*

Definition 4.2. *A transfer matrix $W(s)$ is a J-spectral factor of $\Lambda(s)$ if $W(s)$ is bistable and $\Lambda(s) = W^\sim(s) J W(s)$. Such a factorisation of $\Lambda(s)$ is referred to as a J-spectral factorisation.*

Definition 4.3. *A matrix $W(s)$ is a J-spectral co-factor of a matrix $\Lambda(s)$ if $W(s)$ is bistable and $\Lambda(s) = W(s) J W^\sim(s)$. Such a factorisation of $\Lambda(s)$ is referred to as a J-spectral co-factorisation.*

Theorem 4.1. *A given square, minimal, rational matrix $\Lambda(s)$, having no poles or zeros on the $j\omega$-axis including ∞, is a para-Hermitian matrix if and only if a minimal realization can be represented as*

---

[3] The eigenvalues of an idempotent matrix are either 0 or 1. The eigenvalues of a nilpotent matrix A are all 0 as it satisfies $A^k = 0$ for some positive integer matrix power k.

$$\Lambda = \left[\begin{array}{cc|c} A & R & -B \\ -E & -A^* & C^* \\ \hline C & B^* & D \end{array}\right] \qquad (4.1)$$

where $D = D^*$, $E = E^*$ and $R = R^*$.

Proof. Sufficiency. It is obvious according to Definition 4.1.

Necessity. Since Λ is a para-Hermitian matrix, $D = D^*$. By assumption D is invertible, then using similar arguments as in [32, pp.90–91], Λ^{-1} exists and can be minimally realized as

$$\Lambda^{-1} = \left[\begin{array}{cc|c} A_1 & 0 & B_1 \\ 0 & -A_1^* & -C_1^* \\ \hline C_1 & B_1^* & D^{-1} \end{array}\right]$$

where (A_1, B_1, C_1) is a stable minimal realization. Hence,

$$\Lambda = \left[\begin{array}{cc|c} A_1 - B_1 D C_1 & -B_1 D B_1^* & -B_1 \\ C_1^* D C_1 & -(A_1 - B_1 D C_1)^* & C_1^* \\ \hline C_1 & B_1^* & D \end{array}\right]. \qquad (4.2)$$

This matrix is in the form of (4.1) where $E = E^* = -C_1^* D C_1$ and $R = R^* = -B_1 D B_1^*$. □

Remark 4.1. For the realization of Λ in (4.2), $R = R^*$ is equal to $-B_1 D B_1^*$. However, this is not true for the realization of Λ in (4.1), where $R = R^*$ is in general not equal to $-BDB^*$. A similar argument applies to $E = E^*$.

Remark 4.2. The para-Hermitian transfer matrix characterised in Theorem 4.1 is called *regular*. It says that a regular para-Hermitian transfer matrix Λ realized in the general state-space form

$$\Lambda = \left[\begin{array}{c|c} H_p & B_\Lambda \\ \hline C_\Lambda & D \end{array}\right] \qquad (4.3)$$

can always be transformed into the form of (4.1) after a certain similarity transformation. Here, the A-matrix of Λ is denoted by H_p. Denote the A-matrix of Λ^{-1} by H_z, then

$$H_z = H_p - B_\Lambda D^{-1} C_\Lambda.$$

The eigenvalues of H_p and H_z are, respectively, the poles and zeros of Λ.

Denote $T = \left[\begin{array}{c|c} A_1 & B_1 \\ \hline C_1 & \frac{1}{2}D^{-1} \end{array}\right]$, then the result proposed in [74, Corollary 3.1] can be directly used to find the J-spectral factorisation of Λ. However, as explained before, this will result in a J-spectral factor in terms of A_1, B_1, C_1 and D but not in terms of the original realization in A, R, E, B, C and D. This is good enough for numerical computation, but not for further analysis, as in the case of [160].

The next section is devoted to the elementary characteristics of J-spectral factorisation for regular para-Hermitian transfer matrices realized in the general form (4.3). Section 4.5 is devoted to the J-spectral factorisation of a smaller subset with realization in the form (4.1).

4.4 J-spectral Factorisation of the Full Set

4.4.1 Via Similarity Transformations with Two Matrices

Assume that a para-Hermitian matrix Λ as given in (4.3) is minimal and has no poles or zeros on the $j\omega$-axis including ∞. There always exist nonsingular matrices Δ_p and Δ_z (e.g., via Schur decomposition) such that[4]

$$\Delta_p^{-1} H_p \Delta_p = \begin{bmatrix} ? & 0 \\ ? & A_+ \end{bmatrix} \quad (4.4)$$

and

$$\Delta_z^{-1} H_z \Delta_z = \begin{bmatrix} A_- & ? \\ 0 & ? \end{bmatrix}, \quad (4.5)$$

where A_+ is antistable and A_- is stable (A_+ and A_- have the same dimension).

Lemma 4.1. Λ *described as above has a $J_{p,q}$-spectral factorisation for some unique signature matrix $J_{p,q}$ (where p is the number of the positive eigenvalues of D and q is the number of the negative eigenvalues of D) iff*

$$\Delta = \begin{bmatrix} \Delta_z \begin{bmatrix} I \\ 0 \end{bmatrix} & \Delta_p \begin{bmatrix} 0 \\ I \end{bmatrix} \end{bmatrix} \quad (4.6)$$

is nonsingular. If this condition is satisfied, then a J-spectral factor is formulated as

$$W = \left[\begin{array}{c|c} [I\ 0]\, \Delta^{-1} H_p \Delta \begin{bmatrix} I \\ 0 \end{bmatrix} & [I\ 0]\, \Delta^{-1} B_\Lambda \\ \hline J_{p,q} D_W^{-*} C_\Lambda \Delta \begin{bmatrix} I \\ 0 \end{bmatrix} & D_W \end{array} \right], \quad (4.7)$$

where D_W is a nonsingular solution of $D_W^* J_{p,q} D_W = D$.

Proof. Formulae (4.4) and (4.5) mean that

$$H_p \Delta_p \begin{bmatrix} 0 \\ I \end{bmatrix} = \Delta_p \begin{bmatrix} 0 \\ I \end{bmatrix} A_+$$

and

$$H_z \Delta_z \begin{bmatrix} I \\ 0 \end{bmatrix} = \Delta_z \begin{bmatrix} I \\ 0 \end{bmatrix} A_-.$$

Hence, $\Delta_p \begin{bmatrix} 0 \\ I \end{bmatrix}$ and $\Delta_z \begin{bmatrix} I \\ 0 \end{bmatrix}$ span the antistable eigenspace $\mathcal{M}$ of H_p and the stable eigenspace $\mathcal{M}^\times$ of H_z, respectively. As is well known [12, 32, 39, 74, 119], there exists a J-spectral factorisation iff $\mathcal{M} \cap \mathcal{M}^\times = \{0\}$, which is equivalent to the Δ given in (4.6) being nonsingular.

When this condition holds, there exists a projection P onto $\mathcal{M}^\times$ along $\mathcal{M}$. According to Section 4.2, the projection matrix P is given by

---
[4] The elements denoted by "?" are irrelevant.

$$P = \Delta \begin{bmatrix} I & 0 \\ 0 & 0 \end{bmatrix} \Delta^{-1}. \tag{4.8}$$

With this projection formula, it is easy to formulate a J-spectral factor of Λ [11, 12, 119]. A J-spectral factor is given by

$$W = \left[\begin{array}{c|c} PH_pP & PB_\Lambda \\ \hline J_{p,q}D_W^{-*}C_\Lambda P & D_W \end{array} \right]. \tag{4.9}$$

This realization is not minimal since the A-matrix of W has the same dimension as H_p. Substitute (4.8) into (4.9) and apply a similarity transformation with Δ, then

$$W = \left[\begin{array}{c|c} \Delta^{-1}PH_p\Delta \begin{bmatrix} I & 0 \\ 0 & 0 \end{bmatrix} & \Delta^{-1}PB_\Lambda \\ \hline J_{p,q}D_W^{-*}C_\Lambda\Delta \begin{bmatrix} I & 0 \\ 0 & 0 \end{bmatrix} & D_W \end{array} \right].$$

After removing the unobservable states by deleting the second row and the second column, W becomes

$$W = \left[\begin{array}{c|c} \begin{bmatrix} I & 0 \end{bmatrix} \Delta^{-1}PH_p\Delta \begin{bmatrix} I \\ 0 \end{bmatrix} & \begin{bmatrix} I & 0 \end{bmatrix} \Delta^{-1}PB_\Lambda \\ \hline J_{p,q}D_W^{-*}C_\Lambda\Delta \begin{bmatrix} I \\ 0 \end{bmatrix} & D_W \end{array} \right].$$

Since $\begin{bmatrix} I & 0 \end{bmatrix} \Delta^{-1}P = \begin{bmatrix} I & 0 \end{bmatrix} \Delta^{-1}$, W can be further simplified as given in (4.7). □

4.4.2 Via Similarity Transformations with One Matrix

In general, $\Delta_z \neq \Delta_p$. However, these two can be the same.

Theorem 4.2. *Assume that a para-Hermitian matrix Λ as given in (4.3) is minimal and has no poles or zeros on the $j\omega$-axis including ∞. Then Λ admits a J-spectral factorisation if and only if there exists a nonsingular matrix Δ such that*

$$\Delta^{-1}H_p\Delta = \begin{bmatrix} A^p_- & 0 \\ ? & A^p_+ \end{bmatrix} \tag{4.10}$$

and

$$\Delta^{-1}H_z\Delta = \begin{bmatrix} A^z_- & ? \\ 0 & A^z_+ \end{bmatrix}, \tag{4.11}$$

where A^z_- and A^p_- are stable, and A^z_+ and A^p_+ are antistable. In this case, a J-spectral factor W is as given in (4.7).

Proof. Sufficiency. It is obvious according to Lemma 4.1. In this case, $\Delta_z = \Delta_p = \Delta$. *Necessity.* If there exists a J-spectral factorisation then the Δ given in (4.6) is nonsingular. This Δ does satisfy (4.10) and (4.11).

Since this Δ is the same as that in Lemma 4.1, the J-spectral factor is the same as given in (4.7). □

Remark 4.3. The simultaneous triangularisation of H_p and H_z in (4.10) is of theoretical value. In practice, Δ can be constructed according to (4.6) after two ordered Schur decompositions or similarity transformations.

Remark 4.4. The A-matrix of W is

$$[I\ 0]\,\Delta^{-1}H_p\Delta\begin{bmatrix}I\\0\end{bmatrix} = A_{-}^{p}$$

and that of W^{-1} is

$$[I\ 0]\,\Delta^{-1}H_z\Delta\begin{bmatrix}I\\0\end{bmatrix} = A_{-}^{z}.$$

Remark 4.5. This means a J-spectral factorisation exists if and only if there exists a common similarity transformation to transform H_p (H_z, resp.) into a 2×2 lower (upper, resp.) triangular block matrix with the $(1,1)$-block including all the stable modes of H_p (H_z, resp.). Once the similarity transformation is done, a J-spectral factor can be formulated according to (4.7). If there is no such a similarity transformation, then there is no J-spectral factorisation.

4.5 J-spectral Factorisation of a Smaller Subset

In this section, a subset of the class of para-Hermitian matrices characterised in Theorem 4.1 with the realization of (4.1) is considered. In this case,

$$H_p = \begin{bmatrix} A & R \\ -E & -A^* \end{bmatrix},$$

$$H_z = \begin{bmatrix} A & R \\ -E & -A^* \end{bmatrix} - \begin{bmatrix} -B \\ C^* \end{bmatrix} D^{-1} [C\ B^*] \doteq \begin{bmatrix} A_z & R_z \\ -E_z & -A_z^* \end{bmatrix}.$$

Theorem 4.3. *For a para-Hermitian matrix Λ characterised in Theorem 4.1, assume that: (i) (E, A) is detectable and E is sign definite; (ii) (A_z, R_z) is stabilisable and R_z is sign definite. Then the two ARE*

$$[I\ -L_o]\,H_p\begin{bmatrix}L_o\\I\end{bmatrix} = 0 \qquad (4.12)$$

and

$$[-L_c\ I]\,H_z\begin{bmatrix}I\\L_c\end{bmatrix} = 0 \qquad (4.13)$$

always have unique symmetric solutions L_o and L_c, respectively, such that $\begin{bmatrix} I & -L_o \end{bmatrix} H_p \begin{bmatrix} I \\ 0 \end{bmatrix}$ and $\begin{bmatrix} I & 0 \end{bmatrix} H_z \begin{bmatrix} I \\ L_c \end{bmatrix}$ are stable. In this case, $\Lambda(s)$ has a $J_{p,q}$-spectral factorisation for some unique $J_{p,q}$ (where p is the number of positive eigenvalues of D and q is the number of negative eigenvalues of D) if and only if $\det(I - L_o L_c) \neq 0$. If this condition is satisfied, then one J-spectral factor is formulated as

$$W = \left[\begin{array}{c|c} A + L_o E & B + L_o C^* \\ \hline -J_{p,q} D_W^{-*}(B^* L_c + C)(I - L_o L_c)^{-1} & D_W \end{array} \right],$$

where D_W is nonsingular and $D_W^* J_{p,q} D_W = D$.

Proof. In this case, $\Delta_z = \begin{bmatrix} I & 0 \\ L_c & I \end{bmatrix}$, $\Delta_p = \begin{bmatrix} I & L_o \\ 0 & I \end{bmatrix}$ and $\Delta = \begin{bmatrix} I & L_o \\ L_c & I \end{bmatrix}$. According to Lemma 4.1, there exists a J-spectral factorisation iff Δ is nonsingular, i.e., $\det(I - L_o L_c) \neq 0$. Substitute Δ into (4.7) and apply a similarity transformation with $-(I - L_o L_c)^{-1}$, then

$$W = \left[\begin{array}{c|c} \begin{bmatrix} I & L_o \end{bmatrix} H_p \begin{bmatrix} I \\ L_c \end{bmatrix} (I - L_o L_c)^{-1} & B + L_o C^* \\ \hline -J_{p,q} D_W^{-*}(C + B^* L_c)(I - L_o L_c)^{-1} & D_W \end{array} \right]$$

$$= \left[\begin{array}{c|c} \begin{bmatrix} I & -L_o \end{bmatrix} H_p \begin{bmatrix} I \\ 0 \end{bmatrix} & B + L_o C^* \\ \hline -J_{p,q} D_W^{-*}(C + B^* L_c)(I - L_o L_c)^{-1} & D_W \end{array} \right],$$

where the ARE (4.12) was used. $\square$

A different approach involving two similarity transformations to derive the realization of W is shown below.

Similarity Transformation 1: Stabilisation

Since L_o is the stabilising solution of (4.12), after applying a similarity transformation $\begin{bmatrix} I & L_o \\ 0 & I \end{bmatrix}$, Λ is equal to

$$\Lambda = \left[\begin{array}{cc|c} A + L_o E & 0 & -(B + L_o C^*) \\ -E & -A^* - E L_o & C^* \\ \hline C & B^* + C L_o & D \end{array} \right]. \quad (4.14)$$

Similarity Transformation 2: Factorisation

Since L_c is the stabilising solution of (4.13) and $\det(I - L_c L_o) \neq 0$, the following self-adjoint matrix is well defined:

$$L_{co} = L_c (I - L_o L_c)^{-1} = (I - L_c L_o)^{-1} L_c = L_{co}^*. \quad (4.15)$$

Moreover,

$$I + L_o L_{co} = (I - L_o L_c)^{-1} \quad \text{and} \quad I + L_{co} L_o = (I - L_c L_o)^{-1}$$

are nonsingular. As a result, L_c can be represented as

$$L_c = (I + L_{co}L_o)^{-1}L_{co} = L_{co}(I + L_oL_{co})^{-1},$$

and the ARE (4.13) is equivalent to

$$\begin{bmatrix} -L_{co} & I + L_{co}L_o \end{bmatrix} H_z \begin{bmatrix} I + L_oL_{co} \\ L_{co} \end{bmatrix} = 0, \qquad (4.16)$$

which can be expanded as the following equality using (4.12):

$$L_{co}(A + L_oE) + (A^* + EL_o)L_{co} + E$$
$$= -[L_{co}B + (I + L_{co}L_o)C^*] D^{-1} [B^*L_{co} + C(I + L_oL_{co})]. \qquad (4.17)$$

According to the assumptions, Λ does not have any eigenvalue on the $j\omega$ axis including ∞. Assume p is the number of the positive eigenvalues of D and q is the number of the negative eigenvalues of D, then the equation $D_W^* J_{p,q} D_W = D$ has an invertible solution D_W.

Carrying on another similarity transformation $\begin{bmatrix} I & 0 \\ L_{co} & I \end{bmatrix}$ with respect to (4.14), then

$$\Lambda = \left[\begin{array}{cc|c} A + L_oE & 0 & B + L_oC^* \\ L_{co}(A + L_oE) + (A^* + EL_o)L_{co} + E & -A^* - EL_o & C^* + L_{co}(B + L_oC^*) \\ -C - (B^* + CL_o)L_{co} & B^* + CL_o & D_W^* J_{p,q} D_W \end{array}\right],$$

where an additional similarity transformation $\begin{bmatrix} -I & 0 \\ 0 & I \end{bmatrix}$ was applied. Due to the equality (4.17), the above Λ can be factorised as $\Lambda = W^\sim \cdot J_{p,q} \cdot W$ with

$$W \doteq \left[\begin{array}{c|c} A + L_oE & B + L_oC^* \\ \hline -J_{p,q}^{-1} D_W^{-*} [B^*L_{co} + C(I + L_oL_{co})] & D_W \end{array}\right].$$

Using (4.15), W can be simplified as given in Theorem 4.3. W is bistable because the A-matrix of W is $\begin{bmatrix} I & -L_o \end{bmatrix} H_p \begin{bmatrix} I \\ 0 \end{bmatrix} = A + L_oE$ and the A-matrix of W^{-1} is

$$A + L_oE + (B + L_oC^*)D_W^{-1}J_{p,q}^{-1}D_W^{-*}(B^*L_c + C)(I - L_oL_c)^{-1}$$
$$= \begin{bmatrix} I & -L_o \end{bmatrix} H_z \begin{bmatrix} I \\ L_c \end{bmatrix} (I - L_oL_c)^{-1}$$
$$\sim (I - L_oL_c)^{-1} \begin{bmatrix} I & -L_o \end{bmatrix} H_z \begin{bmatrix} I \\ L_c \end{bmatrix}$$
$$= (I - L_oL_c)^{-1} \left(\begin{bmatrix} I & -L_o \end{bmatrix} H_z \begin{bmatrix} I \\ L_c \end{bmatrix} + L_o \begin{bmatrix} -L_c & I \end{bmatrix} H_z \begin{bmatrix} I \\ L_c \end{bmatrix} \right)$$
$$= \begin{bmatrix} I & 0 \end{bmatrix} H_z \begin{bmatrix} I \\ L_c \end{bmatrix},$$

where "$\sim$" means "similar to" and the ARE (4.13) was used. W is indeed a J-spectral factor of Λ.

Dually to Theorem 4.3, the following theorem holds (with proof omitted):

Theorem 4.4. *For a para-Hermitian matrix Λ characterised in Theorem 4.1, assume that: (i) (A, R) is stabilisable and R is sign definite; (ii) (E_z, A_z) is detectable and E_z is sign definite. Then the two ARE*

$$\begin{bmatrix} -L_c & I \end{bmatrix} H_p \begin{bmatrix} I \\ L_c \end{bmatrix} = 0$$

and

$$\begin{bmatrix} I & -L_o \end{bmatrix} H_z \begin{bmatrix} L_o \\ I \end{bmatrix} = 0$$

always have unique symmetric solutions L_c and L_o, respectively, such that $\begin{bmatrix} I & 0 \end{bmatrix} H_p \begin{bmatrix} I \\ L_c \end{bmatrix}$ and $\begin{bmatrix} I & -L_o \end{bmatrix} H_z \begin{bmatrix} I \\ 0 \end{bmatrix}$ are stable. In this case, $\Lambda(s)$ has a $J_{p,q}$-spectral factorisation for some unique $J_{p,q}$ (where p is the number of positive eigenvalues of D and q is the number of negative eigenvalues of D) if and only if $\det(I - L_o L_c) \neq 0$. If this condition is satisfied, then one J-spectral co-factor is formulated as

$$W(s) = \left[\begin{array}{c|c} A + RL_c & -(I - L_o L_c)^{-1}(B + L_o C^*) D_W^{-*} J_{p,q} \\ \hline B^* L_c + C & D_W \end{array} \right],$$

where D_W is nonsingular and $D_W J_{p,q} D_W^* = D$.

4.6 J-spectral Factorisation of $\Lambda = G^\sim JG$ with Stable G

Here, $J = J^*$ is a signature matrix. Assume that $G(s)$ does not have poles or zeros on the $j\omega$-axis including ∞ and the following realization of $G(s)$ is minimal:

$$G(s) = \left[\begin{array}{c|c} A & B \\ \hline C & D \end{array} \right].$$

This has been discussed extensively [74, 119].

The realization of $\Lambda = G^\sim JG$ is

$$\Lambda = \left[\begin{array}{cc|c} A & 0 & B \\ -C^*JC & -A^* & -C^*JD \\ \hline D^*JC & B^* & D^*JD \end{array} \right].$$

The A-matrix of Λ is

$$H_p = \begin{bmatrix} A & 0 \\ -C^*JC & -A^* \end{bmatrix}$$

and the A-matrix of Λ^{-1} is

$$H_z = \begin{bmatrix} A & 0 \\ -C^*JC & -A^* \end{bmatrix} - \begin{bmatrix} B \\ -C^*JD \end{bmatrix} (D^*JD)^{-1} \begin{bmatrix} D^*JC & B^* \end{bmatrix}.$$

Since A is stable, there is no similarity transformation needed to bring H_p into the form (4.4). This means

4.6 J-spectral Factorisation of $\Lambda = G^\sim JG$ with Stable G

$$\Delta_p = \begin{bmatrix} I & 0 \\ 0 & I \end{bmatrix},$$

which gives a possible Δ in the following form

$$\Delta = \begin{bmatrix} X_1 & 0 \\ X_2 & I \end{bmatrix}$$

to bring H_z into the form (4.11). According to Theorem 4.2, Λ admits a J-spectral factorisation for some unique $J_{p,q}$ (where p is the number of positive eigenvalues of D^*JD and q is the number of negative eigenvalues of D^*JD) iff Δ and, furthermore X_1, are nonsingular. When X_1 is nonsingular, then a further similarity transformation with $\begin{bmatrix} X_1^{-1} & 0 \\ 0 & I \end{bmatrix}$ can be done. This gives another Δ as

$$\Delta = \begin{bmatrix} I & 0 \\ X & I \end{bmatrix}$$

with $X = X_2 X_1^{-1}$. Substitute this Δ into (4.11), then

$$\begin{bmatrix} I & 0 \\ -X & I \end{bmatrix} H_z \begin{bmatrix} I & 0 \\ X & I \end{bmatrix} = \begin{bmatrix} A_-^z & ? \\ 0 & A_+^z \end{bmatrix}.$$

Hence,

$$\begin{bmatrix} -X & I \end{bmatrix} H_z \begin{bmatrix} I \\ X \end{bmatrix} = 0, \tag{4.18}$$

and

$$A_-^z = \begin{bmatrix} I & 0 \end{bmatrix} H_z \begin{bmatrix} I \\ X \end{bmatrix}$$

is stable. Combining these two conditions, the conclusion is that, under the assumptions stated, Λ admits a J-spectral factorisation iff the ARE (4.18) has a stabilising solution X. When this condition holds, the J-spectral factor can easily be found, according to Theorem 4.2, as

$$W = \left[\begin{array}{c|c} \begin{bmatrix} I & 0 \end{bmatrix} \begin{bmatrix} I & 0 \\ -X & I \end{bmatrix} H_p \begin{bmatrix} I & 0 \\ X & I \end{bmatrix} \begin{bmatrix} I \\ 0 \end{bmatrix} & \begin{bmatrix} I & 0 \end{bmatrix} \begin{bmatrix} I & 0 \\ -X & I \end{bmatrix} \begin{bmatrix} B \\ -C^* JD \end{bmatrix} \\ \hline J_{p,q} D_W^{-*} \begin{bmatrix} D^*JC & B^* \end{bmatrix} \begin{bmatrix} I & 0 \\ X & I \end{bmatrix} \begin{bmatrix} I \\ 0 \end{bmatrix} & D_W \end{array} \right]$$

$$= \left[\begin{array}{c|c} A & B \\ \hline J_{p,q} D_W^{-*}(D^*JC + B^*X) & D_W \end{array} \right],$$

where D_W is nonsingular and $D_W^* J_{p,q} D_W = D^*JD$. This is consistent with the known results.

4.7 Numerical Examples

4.7.1 $\Lambda(s) = \begin{bmatrix} 0 & \frac{s-1}{s+1} \\ \frac{s+1}{s-1} & 0 \end{bmatrix}$

This example has been considered in [119]. A minimal realization of Λ is given by

$$\Lambda = \left[\begin{array}{cc|cc} 1 & 0 & 1 & 0 \\ 0 & -1 & 0 & -2 \\ \hline 0 & 1 & 0 & 1 \\ 2 & 0 & 1 & 0 \end{array}\right].$$

This gives

$$H_p = \begin{bmatrix} 1 & 0 \\ 0 & -1 \end{bmatrix}$$

and

$$H_z = \begin{bmatrix} 1 & 0 \\ 0 & -1 \end{bmatrix} - \begin{bmatrix} 1 & 0 \\ 0 & -2 \end{bmatrix}\begin{bmatrix} 0 & 1 \\ 1 & 0 \end{bmatrix}^{-1}\begin{bmatrix} 0 & 1 \\ 2 & 0 \end{bmatrix} = \begin{bmatrix} -1 & 0 \\ 0 & 1 \end{bmatrix}.$$

Apparently, there does not exist a common similarity transformation to make the (1, 1)-elements of H_p and H_z all stable. Hence, this Λ does not admit a J-spectral factorisation.

4.7.2 $\Lambda(s) = \begin{bmatrix} -\frac{s^2-4}{s^2-1} & 0 \\ 0 & \frac{s^2-1}{s^2-4} \end{bmatrix}$

A minimal realization of Λ is

$$\Lambda = \left[\begin{array}{cccc|cc} 0 & \frac{1}{2} & 0 & 0 & 1 & 0 \\ 2 & 0 & 0 & 0 & 0 & 0 \\ 0 & 0 & 0 & 2 & 0 & 1 \\ 0 & 0 & 2 & 0 & 0 & 0 \\ \hline 0 & \frac{3}{2} & 0 & 0 & -1 & 0 \\ 0 & 0 & 0 & \frac{3}{2} & 0 & 1 \end{array}\right].$$

This gives the A-matrices of Λ and Λ^{-1}, respectively, as

$$H_p = \begin{bmatrix} 0 & \frac{1}{2} & 0 & 0 \\ 2 & 0 & 0 & 0 \\ 0 & 0 & 0 & 2 \\ 0 & 0 & 2 & 0 \end{bmatrix} \quad \text{and} \quad H_z = \begin{bmatrix} 0 & 2 & 0 & 0 \\ 2 & 0 & 0 & 0 \\ 0 & 0 & 0 & \frac{1}{2} \\ 0 & 0 & 2 & 0 \end{bmatrix}.$$

By doing similarity transformations, it is easy to transform H_p and H_z into a lower (upper, resp.) triangular matrix with the first two diagonal elements being negative. In order to bring H_p into a lower triangular matrix, two steps are used. The first step is to bring it into an upper triangular matrix and the second step is to group the stable modes, as shown below.

$$\begin{bmatrix} 0 & \frac{1}{2} & 0 & 0 \\ 2 & 0 & 0 & 0 \\ 0 & 0 & 0 & 2 \\ 0 & 0 & 2 & 0 \end{bmatrix}$$

$$\downarrow \quad \text{with} \quad \Delta_{p1} = \begin{bmatrix} 1 & 0 & 0 & 0 \\ 2 & 1 & 0 & 0 \\ 0 & 0 & 1 & 0 \\ 0 & 0 & 1 & 1 \end{bmatrix}$$

$$\begin{bmatrix} 1 & \frac{1}{2} & 0 & 0 \\ 0 & -1 & 0 & 0 \\ 0 & 0 & 2 & 2 \\ 0 & 0 & 0 & -2 \end{bmatrix}$$

$$\downarrow \quad \text{with} \quad \Delta_{p2} = \begin{bmatrix} 0 & 0 & 0 & 1 \\ 0 & 1 & 0 & 0 \\ 0 & 0 & 1 & 0 \\ 1 & 0 & 0 & 0 \end{bmatrix}$$

$$\begin{bmatrix} -2 & 0 & 0 & 0 \\ 0 & -1 & 0 & 0 \\ 2 & 0 & 2 & 0 \\ 0 & \frac{1}{2} & 0 & 1 \end{bmatrix}.$$

This is a lower triangular matrix, to which H_p is similarly transformed with

$$\Delta_p = \Delta_{p1}\Delta_{p2} = \begin{bmatrix} 0 & 0 & 0 & 1 \\ 0 & 1 & 0 & 2 \\ 0 & 0 & 1 & 0 \\ 1 & 0 & 1 & 0 \end{bmatrix}.$$

In general, a third step is needed to make the nonzero elements, if any, in the upper-right area zero.

Similarly, H_z is transformed into an upper triangular matrix

$$\Delta_z^{-1} H_z \Delta_z = \begin{bmatrix} -1 & 0 & 2 & 0 \\ 0 & -2 & 0 & 2 \\ 0 & 0 & 1 & 0 \\ 0 & 0 & 0 & 2 \end{bmatrix}, \quad \text{with} \quad \Delta_z = \begin{bmatrix} 0 & -1 & 0 & 1 \\ 0 & 1 & 0 & 0 \\ -\frac{1}{2} & 0 & 1 & 0 \\ 1 & 0 & 0 & 0 \end{bmatrix}.$$

As a result, Δ can be obtained by combining the first two columns of Δ_z and the last two columns of Δ_p as

$$\Delta = \begin{bmatrix} 0 & -1 & 0 & 1 \\ 0 & 1 & 0 & 2 \\ -\frac{1}{2} & 0 & 1 & 0 \\ 1 & 0 & 1 & 0 \end{bmatrix}.$$

This Δ is nonsingular and, hence, there exists a J-spectral factorisation.

The D-matrix $D = \begin{bmatrix} -1 & 0 \\ 0 & 1 \end{bmatrix}$ of Λ has one positive eigenvalue and a negative eigenvalue. This gives

$$J_{p,q} = \begin{bmatrix} 1 & 0 \\ 0 & -1 \end{bmatrix} \quad \text{and} \quad D_W = \begin{bmatrix} 0 & 1 \\ 1 & 0 \end{bmatrix}$$

such that $D_W^* J_{p,q} D_W = D$. According to (4.7), a J-spectral factor of Λ is found as

$$W = \left[\begin{array}{cc|cc} -2 & 0 & 0 & -\frac{2}{3} \\ 0 & -1 & -\frac{2}{3} & 0 \\ \hline \frac{3}{2} & 0 & 0 & 1 \\ 0 & -\frac{3}{2} & 1 & 0 \end{array}\right],$$

which can be described in transfer matrix as

$$W(s) = \begin{bmatrix} 0 & \frac{s+1}{s+2} \\ \frac{s+2}{s+1} & 0 \end{bmatrix}.$$

It is worth noting that a corresponding J-spectral factor for

$$\bar{\Lambda}(s) = -\Lambda(s) = \begin{bmatrix} \frac{s^2-4}{s^2-1} & 0 \\ 0 & -\frac{s^2-1}{s^2-4} \end{bmatrix}$$

with $J_{p,q} = \begin{bmatrix} 1 & 0 \\ 0 & -1 \end{bmatrix}$ is

$$\bar{W}(s) = \begin{bmatrix} 0 & 1 \\ 1 & 0 \end{bmatrix} W(s) = \begin{bmatrix} \frac{s+2}{s+1} & 0 \\ 0 & \frac{s+1}{s+2} \end{bmatrix}$$

because

$$\begin{bmatrix} 0 & 1 \\ 1 & 0 \end{bmatrix} J_{p,q} \begin{bmatrix} 0 & 1 \\ 1 & 0 \end{bmatrix} = -J_{p,q}.$$

4.8 Summary

A class of regular invertible para-Hermitian transfer matrices is characterised and then the J-spectral factorisation of transfer matrices is studied. A transfer matrix Λ in this class admits a J-spectral factorisation if and only if there exists a common nonsingular matrix to similarly transform the A-matrices of Λ and Λ^{-1}, resp., into 2×2 lower (upper, resp.) triangular block matrices with the $(1,1)$-block including all the stable modes of Λ (Λ^{-1}, resp.). The resulting J-spectral factor is formulated in terms of the original realization of Λ. When the transfer matrix meets additional conditions, there exists a J-spectral factorisation if and only if a coupling condition related to the stabilising solutions of two AREs holds. A well-studied case with $\Lambda = G^\sim JG$ with a stable G is revisited and two numerical examples are shown.

5

The Delay-type Nehari Problem

This chapter[1] generalises the frequency-domain results for the delay-type Nehari problem in the stable case and the unstable case. It also extends the solution of the conventional (delay-free) Nehari problem to the delay-type Nehari problem. The solvability condition of the delay-type Nehari problem is formulated in terms of the nonsingularity of a delay-dependent matrix. The optimal value γ_{opt} is the maximal $\gamma \in [0, \infty)$ such that this matrix becomes singular when γ decreases from ∞. All suboptimal compensators are parameterised in a transparent structure incorporating a modified Smith predictor. The arguments about J-spectral factorisation developed in the previous chapter are crucial to derivation of the results.

5.1 Introduction

The H^∞ control of processes with delay(s) has been an active research area since the mid 1980s. It is well known that a large class of H^∞ control problems can be reduced to Nehari problems [32]. This is still true in the case of systems with delay(s), where the simplified problem is a delay-type Nehari problem. Some papers [29, 182] were devoted to calculation of the infimum of the delay-type Nehari problem in the stable case. It was shown in [182] that this problem in the stable case is equivalent to calculating an $L_2[0, h]$-induced norm. However, for the unstable case, it becomes much more complicated. Tadmor [132] presented a state-space solution to this problem in the unstable case, in which a differential/algebraic matrix Riccati equation-based method was used. The optimal value relies on the solution of a differential Riccati equation. Although the expositions are very elegant, the suboptimal solution is too complicated and the structure is not transparent.

Motivated by the idea of Meinsma and Zwart [80], this chapter presents a frequency-domain solution to the delay-type Nehari problem in the unstable

---

[1] Portions reprinted from [158, 160], with permission from Elsevier.

case. The optimal value γ_{opt} of the delay-type Nehari problem is formulated in a simple and clear way: it is the maximal γ such that a delay-dependent matrix becomes singular when γ decreases from $+\infty$ to 0. Hence, one need no longer solve a differential Riccati equation. All the suboptimal compensators are formulated in a transparent structure incorporating a modified Smith predictor (but not of the original plant).

5.2 Problem Statement (NP$_h$)

The Delay-type Nehari Problem (NP$_h$) can be described as follows:

Given a minimal state-space realization

$$G_\beta \doteq \left[\begin{array}{c|c} A & B \\ \hline -C & 0 \end{array}\right], \qquad (5.1)$$

which is not necessarily stable and $h \geq 0$, characterise the optimal value

$$\gamma_{opt} = \inf\{\|G_\beta(s) + e^{-sh}K(s)\|_{L_\infty} : K(s) \in H^\infty\} \qquad (5.2)$$

and for a given $\gamma > \gamma_{opt}$, parameterise the suboptimal set of proper $K \in H^\infty$ such that

$$\|G_\beta(s) + e^{-sh}K(s)\|_{L_\infty} < \gamma. \qquad (5.3)$$

A similar Nehari problem with distributed delays, which can be reformulated as $\|e^{-sh}G_\beta(s) + K(s)\|_\infty < \gamma$ in the single delay case, was studied in [130]. Although this question looks similar to (5.3) in form, they are entirely different in essence, and the problem studied here and in [132] is much more complicated.

The compensator K is, in general, infinite-dimensional. K is called "proper" if there exists a real number α such that

$$\sup_{\operatorname{Re} s > \alpha} \|K(s)\| < \infty,$$

i.e., K is bounded on the right half-plane where $\operatorname{Re} s > \alpha$. K is called "strictly proper" if

$$\lim_{\operatorname{Re} s > \alpha,\, |s| \to +\infty} K(s) = 0.$$

Weiss [151] showed that the properness of K implies the existence of a causal input–output operator with transfer function K.

5.3 Solution to the NP_h

It is well known [36] that this problem is solvable iff

$$\gamma > \gamma_{opt} \doteq \left\|\Gamma_{e^{sh}G_\beta}\right\|, \tag{5.4}$$

where Γ denotes the Hankel operator. The symbol $e^{sh}G_\beta$ is noncausal and, possibly, unstable. It is easy to see from (5.2) that $\gamma_{opt} \le \|G_\beta(s)\|_{L_\infty}$ because at least K can be chosen as 0. It can be seen from (5.4) that $\gamma_{opt} \ge \|\Gamma_{G_\beta}\|$. Hence, the optimal value γ_{opt} satisfies the following inequality:

$$\|\Gamma_{G_\beta}\| \le \gamma_{opt} \le \|G_\beta(s)\|_{L_\infty}. \tag{5.5}$$

When $\gamma \le \|G_\beta(s)\|_{L_\infty}$, the matrix

$$H = \begin{bmatrix} A & \gamma^{-2}BB^* \\ -C^*C & -A^* \end{bmatrix}$$

defined in (3.12) has at least one pair of eigenvalues on the $j\omega$-axis, according to the bounded real lemma [180].

For a given minimally-realized transfer matrix $G_\beta = \left[\begin{array}{c|c} A & B \\ \hline -C & 0 \end{array}\right]$ having neither $j\omega$-axis zero nor $j\omega$-axis pole, define the following two Hamiltonian matrices:

$$H_c = \begin{bmatrix} A & \gamma^{-2}BB^* \\ 0 & -A^* \end{bmatrix}, \qquad H_o = \begin{bmatrix} A & 0 \\ -C^*C & -A^* \end{bmatrix}.$$

Then the ARE

$$\begin{bmatrix} -L_c & I \end{bmatrix} H_c \begin{bmatrix} I \\ L_c \end{bmatrix} = 0 \tag{5.6}$$

and

$$\begin{bmatrix} I & -L_o \end{bmatrix} H_o \begin{bmatrix} L_o \\ I \end{bmatrix} = 0 \tag{5.7}$$

always have unique solutions $L_c \le 0$ and $L_o \le 0$ such that $A + \gamma^{-2}BB^*L_c = \begin{bmatrix} I & 0 \end{bmatrix} H_c \begin{bmatrix} I \\ L_c \end{bmatrix}$ and $A + L_o C^*C = \begin{bmatrix} I & -L_o \end{bmatrix} H_o \begin{bmatrix} I \\ 0 \end{bmatrix}$ are stable.

Theorem 5.1. (Delay-type Nehari problem) *For a given minimally-realized transfer matrix* $G_\beta = \left[\begin{array}{c|c} A & B \\ \hline -C & 0 \end{array}\right]$ *having neither $j\omega$-axis zero nor $j\omega$-axis pole, the optimal value γ_{opt} of the delay-type Nehari problem (5.3) is*

$$\gamma_{opt} = \max\{\gamma : \det \hat{\Sigma}_{22} = 0\},$$

where

$$\hat{\Sigma}_{22} = \begin{bmatrix} -L_c & I \end{bmatrix} \Sigma \begin{bmatrix} L_o \\ I \end{bmatrix}. \tag{5.8}$$

Furthermore, for a given $\gamma > \gamma_{opt}$, all $K(s) \in H^\infty$ satisfying (5.3) can be parameterised as

5 The Delay-type Nehari Problem

$$K = \mathcal{H}_r(\begin{bmatrix} I & 0 \\ Z & I \end{bmatrix} W^{-1}, Q) \qquad (5.9)$$

where $\|Q(s)\|_{H^\infty} < \gamma$ is a free parameter and

$$Z = -\pi_h\{\mathcal{F}_u(\begin{bmatrix} G_\beta & I \\ I & 0 \end{bmatrix}, \gamma^{-2} G_{\tilde{\beta}})\}, \qquad (5.10)$$

$$W^{-1} = \begin{bmatrix} A + \gamma^{-2}BB^*L_c & \hat{\Sigma}_{22}^{-*}(\Sigma_{12}^* + L_o\Sigma_{11}^*)C^* & -\hat{\Sigma}_{22}^{-*}B \\ -C & I & 0 \\ \gamma^{-2}B^*(\Sigma_{21}^* - \Sigma_{11}^*L_c) & 0 & I \end{bmatrix}. \qquad (5.11)$$

Remark 5.1. It can be shown that $\hat{\Sigma}_{22}^* = \begin{bmatrix} I & -L_o \end{bmatrix} \Sigma^{-1} \begin{bmatrix} I \\ L_c \end{bmatrix}$.

The structure of K is shown in Figure 5.1. It consists of an infinite-dimensional block Z, which is a finite-impulse-response (FIR) block (*i.e.*, a modified Smith predictor), a rational block W^{-1} and a free parameter Q.

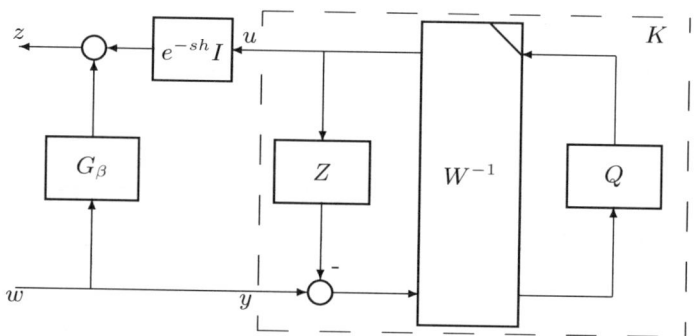

Figure 5.1. Representation of the NP$_h$ as a block diagram

5.4 Proof

The NP$_h$ problem (5.3) can be associated with the following system using the right chain-scattering representation [58]:

$$\begin{bmatrix} z \\ w \end{bmatrix} = \begin{bmatrix} e^{-sh}I & G_\beta(s) \\ 0 & I \end{bmatrix} \begin{bmatrix} u \\ y \end{bmatrix} \doteq G(s) \begin{bmatrix} u \\ y \end{bmatrix},$$

$$u = Ky.$$

The corresponding closed-loop transfer matrix is

$$T_{zw} = \mathcal{H}_r(G, K) = G_\beta + e^{-sh}K.$$

5.4 Proof

It is well known [39, 58, 80] that the H^∞ control problem

$$\|\mathcal{H}_r(G, K)\|_\infty < \gamma$$

is equivalent to $G^\sim J_\gamma G$ having a J-spectral factor V such that the $(2,2)$-block of GV^{-1} is bistable. Hence, in this proof, effort is devoted to characterising the conditions to meet these requirements. Here,

$$G^\sim J_\gamma G = \begin{bmatrix} I & e^{sh}G_\beta \\ e^{-sh}G_{\tilde\beta} & G_{\tilde\beta}G_\beta - \gamma^2 I \end{bmatrix}.$$

The underlying idea here is the *Meinsma–Zwart idea*: to find a unimodular matrix to equivalently rationalise the system and then to find the J-spectral factorisation of the rationalised system, proposed in [80] for a 2-block problem. The result there was obtained for a stable G and cannot be directly used here because G here is not necessarily stable. Some ideas from there are borrowed, but only involving an elementary tool (the similarity transformation) to find a realization of the rationalised system.

Since Z given in (5.10) is a stable FIR block and $\begin{bmatrix} I & 0 \\ Z & I \end{bmatrix}$ is bistable, the following transformation does not affect the solvability of the original problem [58, 80]:

$$\Theta \doteq \begin{bmatrix} I & Z^\sim \\ 0 & I \end{bmatrix} G^\sim J_\gamma G \begin{bmatrix} I & 0 \\ Z & I \end{bmatrix}.$$

Once the J-spectral factor of Θ is obtained, the J-spectral factor of $G^\sim J_\gamma G$ is obtained.

Z given in (5.10) can be decomposed into two parts:

$$Z(s) = F_0(s) + F_1(s)e^{-sh}, \tag{5.12}$$

where

$$F_1 = \mathcal{F}_u(\begin{bmatrix} G_\beta & I \\ I & 0 \end{bmatrix}, \gamma^{-2}G_{\tilde\beta}). \tag{5.13}$$

This eliminates the delay term in

$$\Theta_{21} = e^{-sh}G_{\tilde\beta} + \left(G_{\tilde\beta}G_\beta - \gamma^2 I\right) Z$$

and rationalises Θ as

$$\Theta = \begin{bmatrix} \gamma^2 \left(\gamma^2 I - G_\beta G_{\tilde\beta}\right)^{-1} + F_0^\sim \left(G_{\tilde\beta}G_\beta - \gamma^2 I\right) F_0 & F_0^\sim \left(G_{\tilde\beta}G_\beta - \gamma^2 I\right) \\ \left(G_{\tilde\beta}G_\beta - \gamma^2 I\right) F_0 & G_{\tilde\beta}G_\beta - \gamma^2 I \end{bmatrix} \tag{5.14}$$

Substituting (5.1) into (5.13) and using the formula in Subsection 3.3.1, then F_1 can be obtained as

$$F_1 = \left[\begin{array}{cc|c} A & \gamma^{-2}BB^* & 0 \\ -C^*C & -A^* & C^* \\ \hline 0 & \gamma^{-2}B^* & 0 \end{array}\right]. \tag{5.15}$$

F_0 can be obtained, according to the definition of the operator π_h, as

$$F_0 = \left[\begin{array}{cc|c} A & \gamma^{-2}BB^* & 0 \\ -C^*C & -A^* & C^* \\ \hline \gamma^{-2}B^*\Sigma_{21}^* & -\gamma^{-2}B^*\Sigma_{11}^* & 0 \end{array}\right]. \tag{5.16}$$

Substituting (5.1), (5.15) and (5.16) into (5.14), then Θ^{-1} can be obtained, after tedious matrix manipulations (see Section 5.6), as

$$\Theta^{-1} = \left[\begin{array}{cc|cc} A & \gamma^{-2}BB^* & 0 & \gamma^{-2}\Sigma_{11}B \\ 0 & -A^* & -C^* & \gamma^{-2}\Sigma_{21}B \\ -C & 0 & I & 0 \\ \hline \gamma^{-2}B^*\Sigma_{21}^* & -\gamma^{-2}B^*\Sigma_{11}^* & 0 & -\gamma^{-2}I \end{array}\right].$$

The A-matrix of Θ is $\Sigma H_o \Sigma^{-1}$. Θ^{-1} is in the form of a regular para-Hermitian matrix (4.1) and a J-spectral co-factor W_0^{-1} of Θ^{-1} can be found using similar arguments to those in Chapter 4 (see Section 5.7) as

$$W_0^{-1} = \left[\begin{array}{c|cc} A + \gamma^{-2}BB^*L_c & \hat{\Sigma}_{22}^{-*}(\Sigma_{12}^* + L_o\Sigma_{11}^*)C^* & -\hat{\Sigma}_{22}^{-*}B\gamma^{-1} \\ -C & I & 0 \\ \hline \gamma^{-2}B^*(\Sigma_{21}^* - \Sigma_{11}^*L_c) & 0 & \gamma^{-1}I \end{array}\right],$$

where $\hat{\Sigma}_{22}$ is given in (5.8). Hence, the following equality is obtained:

$$G^\sim J_\gamma G = \begin{bmatrix} I & -Z^\sim \\ 0 & I \end{bmatrix} W_0^\sim J W_0 \begin{bmatrix} I & 0 \\ -Z & I \end{bmatrix}.$$

Since W_0 and $\begin{bmatrix} I & 0 \\ -Z & I \end{bmatrix}$ are all bistable, $W_0 \begin{bmatrix} I & 0 \\ -Z & I \end{bmatrix}$ is a J-spectral factor of $G^\sim J_\gamma G$. This means that any K in the form

$$K = \mathcal{H}_r(\begin{bmatrix} I & 0 \\ Z & I \end{bmatrix} W^{-1}, Q),$$

where

$$W^{-1} = \left[\begin{array}{c|cc} A + \gamma^{-2}BB^*L_c & \hat{\Sigma}_{22}^{-*}(\Sigma_{12}^* + L_o\Sigma_{11}^*)C^* & -\hat{\Sigma}_{22}^{-*}B \\ -C & I & 0 \\ \hline \gamma^{-2}B^*(\Sigma_{21}^* - \Sigma_{11}^*L_c) & 0 & I \end{array}\right],$$

and $\|Q(s)\|_{H^\infty} < \gamma$ is a free parameter, satisfies

$$\|G_\beta(s) + e^{-sh}K(s)\|_{L^\infty} < \gamma.$$

In order to guarantee $K(s) \in H^\infty$, bistability of the $(2,2)$-block of the matrix

$$\Pi = \begin{bmatrix} \Pi_{11} & \Pi_{12} \\ \Pi_{21} & \Pi_{22} \end{bmatrix} \doteq G \begin{bmatrix} I & 0 \\ Z & I \end{bmatrix} W^{-1},$$

i.e., $\Pi_{22} = \begin{bmatrix} Z & I \end{bmatrix} W^{-1} \begin{bmatrix} 0 \\ I \end{bmatrix}$, is required. Using similar arguments to those of Meinsma and Zwart [80], the bistability of Π_{22} is equivalent to the existence of the J-spectral co-factorisation of Θ^{-1} (equivalently, the nonsingularity of $\hat{\Sigma}_{22}$) not only for γ but also for any number larger than γ. The minimal γ which satisfies this condition is γ_{opt}. When this condition is satisfied, the compensator is parameterised as given in (5.9). This completes the proof.

5.5 Special Cases

Three special cases are considered in this section.

5.5.1 The Stable Case

If A is stable, then $L_o = 0$, $L_c = 0$ and thus[2]

$$\hat{\Sigma}_{22} = \begin{bmatrix} 0 & I \end{bmatrix} \Sigma \begin{bmatrix} 0 \\ I \end{bmatrix} = \Sigma_{22}.$$

The optimal γ becomes

$$\gamma_{opt} = \gamma_h = \max\{\gamma : \det \Sigma_{22} = 0\}.$$

This coincides with the Zhou–Khargonekar formula [55] as given in [182].

Corollary 5.1. *For a minimally-realized stable transfer matrix $G_\beta = \left[\begin{array}{c|c} A & B \\ \hline -C & 0 \end{array}\right]$ having neither $j\omega$-axis zero nor $j\omega$-axis pole, the delay-type Nehari problem (5.3) is solvable iff $\gamma > \gamma_h$ (or equivalently, Σ_{22} is nonsingular not only for γ but also for any number larger than γ). Furthermore, if this condition holds, then K is parameterised as (5.9) but*

$$W^{-1} = \left[\begin{array}{c|cc} A & \Sigma_{22}^{-*}\Sigma_{12}^* C^* & -\Sigma_{22}^{-*}B \\ \hline -C & I & 0 \\ \gamma^{-2}B^*\Sigma_{21}^* & 0 & I \end{array}\right].$$

5.5.2 The Conventional Nehari Problem

If delay $h = 0$, then $\Sigma = I$ and thus

$$\hat{\Sigma}_{22} = I - L_o L_c.$$

The condition reduces to the nonsingularity of $I - L_o L_c$ for any $\gamma > \gamma_{opt}$. In this case,

$$\gamma_{opt} = \max\{\gamma : \det(I - L_o L_c) = 0\}.$$

Corollary 5.2. *For a minimally-realized transfer matrix $G_\beta = \left[\begin{array}{c|c} A & B \\ \hline -C & 0 \end{array}\right]$ having neither $j\omega$-axis zero nor $j\omega$-axis pole, the delay-free Nehari problem (to find $K \in H^\infty$ such that $\|G_\beta(s) + K(s)\|_{L_\infty} < \gamma$) is solvable iff*

$$\gamma > \max\{\gamma : \det(I - L_o L_c) = 0\}.$$

Furthermore, if this condition holds, then K is parameterised as

$$K = \mathcal{H}_r\left(W^{-1},\ Q\right),$$

---

[2] This is the reason for the notation $\hat{\Sigma}_{22}$.

where

$$W^{-1} = \left[\begin{array}{c|cc} A+\gamma^{-2}BB^*L_c & (I-L_oL_c)^{-1}L_oC^* & -(I-L_oL_c)^{-1}B \\ -C & I & 0 \\ -\gamma^{-2}B^*L_c & 0 & I \end{array}\right]$$

and $\|Q(s)\|_{H^\infty} < \gamma$ is a free parameter.

Remark 5.2. This is an alternative solution to the well-known Nehari problem which has been addressed extensively [32, 39]. The A-matrix A is not split here. In common situations, it was handled by modal decomposition [39], and the A-matrix A was split into a stable part and an anti-stable part.

5.5.3 The Conventional Nehari Problem with Stable A

If delay $h = 0$ and A is stable, then $L_o = 0$, $L_c = 0$, $\Sigma = I$ and thus

$$\hat{\Sigma}_{22} = I.$$

The condition is always satisfied for any $\gamma \geq 0$. K is then parameterised as

$$K = \mathcal{H}_r\left(W^{-1},\ Q\right),$$

where

$$W^{-1} = \left[\begin{array}{c|cc} A & 0 & B \\ \hline C & I & 0 \\ 0 & 0 & I \end{array}\right] = \begin{bmatrix} I & -G_\beta \\ 0 & I \end{bmatrix}.$$

This is obvious. For the stable delay-free Nehari problem, the solution is apparently

$$K = -G_\beta + Q$$

for any $\gamma > 0$, where $\|Q(s)\|_{H^\infty} < \gamma$ is a free parameter. In this case, $\gamma_{opt} = 0$.

5.6 Realizations of Θ^{-1} and Θ

This section is devoted to deriving realizations of Θ^{-1} and Θ.

Although $\begin{bmatrix} I & F_{\tilde o} \\ 0 & I \end{bmatrix}$ is unstable, Θ can be calculated as follows without changing the result after a series of simplifications (only for calculation, not for implementation):

$$\Theta = \begin{bmatrix} I & F_{\tilde o} \\ 0 & I \end{bmatrix} \begin{bmatrix} \gamma^2 I\left(\gamma^2 I - G_\beta G_{\tilde \beta}\right)^{-1} & 0 \\ 0 & G_{\tilde \beta}G_\beta - \gamma^2 I \end{bmatrix} \begin{bmatrix} I & 0 \\ F_0 & I \end{bmatrix} \quad (5.17)$$

After substituting the realizations of G_β and F_0 into (5.17), Θ can be found as follows. However, this realization is not minimal and has to be minimised, using the state-space operations and the properties of Σ given in Chapter 3.

5.6 Realizations of Θ^{-1} and Θ

$$\Theta = \left[\begin{array}{cccccccc|cc}
A & \gamma^{-2}BB^* & 0 & 0 & 0 & 0 & \gamma^{-2}BB^*\Sigma_{21}^* & -\gamma^{-2}BB^*\Sigma_{11}^* & 0 & 0 \\
-C^*C & -A^* & 0 & 0 & 0 & 0 & 0 & 0 & -C^* & 0 \\
0 & 0 & A & 0 & 0 & 0 & -\gamma^{-2}\Sigma_{21}BB^*\Sigma_{21}^* & \gamma^{-2}\Sigma_{21}BB^*\Sigma_{11}^* & 0 & B \\
0 & 0 & -C^*C & -A^* & 0 & 0 & \gamma^{-2}\Sigma_{11}BB^*\Sigma_{21}^* & -\gamma^{-2}\Sigma_{11}BB^*\Sigma_{11}^* & 0 & 0 \\
0 & 0 & 0 & 0 & A & -\gamma^{-2}BB^* & A & \gamma^{-2}BB^* & 0 & -\Sigma_{21}B \\
0 & 0 & 0 & 0 & -C^*C & -A^* & -C^*C & -A^* & 0 & \Sigma_{11}B \\
\hline
C & 0 & 0 & 0 & 0 & 0 & 0 & 0 & I & 0 \\
0 & 0 & 0 & 0 & -B^*\Sigma_{21}^* & B^*\Sigma_{11}^* & -B^*\Sigma_{21}^* & B^*\Sigma_{11}^* & 0 & -\gamma^2 I
\end{array}\right]$$

$$= \left[\begin{array}{cccccccc|cc}
A & \gamma^{-2}BB^* & 0 & 0 & 0 & 0 & \gamma^{-2}BB^*\Sigma_{21}^* & -\gamma^{-2}BB^*\Sigma_{11}^* & 0 & 0 \\
-C^*C & -A^* & 0 & 0 & 0 & 0 & 0 & 0 & -C^* & 0 \\
0 & 0 & A & 0 & 0 & 0 & -\gamma^{-2}\Sigma_{21}BB^*\Sigma_{21}^* & \gamma^{-2}\Sigma_{21}BB^*\Sigma_{11}^* & 0 & B \\
0 & 0 & -C^*C & -A^* & 0 & 0 & \gamma^{-2}\Sigma_{11}BB^*\Sigma_{21}^* & -\gamma^{-2}\Sigma_{11}BB^*\Sigma_{11}^* & 0 & 0 \\
0 & 0 & 0 & 0 & A & -\gamma^{-2}BB^* & A & \gamma^{-2}BB^* & -C^* & -\Sigma_{21}B \\
0 & 0 & 0 & 0 & -C^*C & -A^* & -C^*C & -A^* & 0 & \Sigma_{11}B \\
\hline
C & 0 & 0 & 0 & 0 & 0 & 0 & 0 & I & 0 \\
0 & 0 & 0 & 0 & -B^*\Sigma_{21}^* & B^*\Sigma_{11}^* & -B^*\Sigma_{21}^* & B^*\Sigma_{11}^* & 0 & -\gamma^2 I
\end{array}\right]$$

$$= \left[\begin{array}{cccccc|cc}
A & 0 & 0 & 0 & \gamma^{-2}BB^*\Sigma_{21}^* & -\gamma^{-2}BB^*\Sigma_{11}^* & B & 0 \\
-C^*C & -A^* & 0 & 0 & 0 & 0 & 0 & 0 \\
0 & 0 & \gamma^{-2}\Sigma_{21}BB^* & 0 & -\gamma^{-2}\Sigma_{21}BB^*\Sigma_{21}^* & \gamma^{-2}\Sigma_{21}BB^*\Sigma_{11}^* & -\Sigma_{21}B & -C^* \\
0 & 0 & -\gamma^{-2}\Sigma_{11}BB^* & -\gamma^{-2}BB^* & \gamma^{-2}\Sigma_{11}BB^*\Sigma_{21}^* & -\gamma^{-2}\Sigma_{11}BB^*\Sigma_{11}^* & \Sigma_{11}B & 0 \\
0 & 0 & 0 & 0 & A & \gamma^{-2}BB^* & 0 & 0 \\
0 & 0 & 0 & 0 & -C^*C & -A^* & 0 & -C^* \\
\hline
0 & 0 & 0 & 0 & -B^*\Sigma_{21}^* & B^*\Sigma_{11}^* & I & 0 \\
0 & 0 & B^* & 0 & -B^*\Sigma_{21}^* & B^*\Sigma_{11}^* & 0 & -\gamma^2 I
\end{array}\right]$$

In order to further simplify Θ, invert it:

$$\Theta^{-1} = \left[\begin{array}{cccccc|cc} A & \gamma^{-2}BB^* & 0 & 0 & 0 & 0 & 0 & \gamma^{-2}B \\ -C^*C & -A^* & 0 & 0 & 0 & 0 & 0 & 0 \\ 0 & 0 & -A^* & 0 & 0 & 0 & C^* & -\gamma^{-2}\Sigma_{21}B \\ 0 & 0 & -\gamma^{-2}BB^* & A & 0 & 0 & 0 & \gamma^{-2}\Sigma_{11}B \\ 0 & 0 & 0 & 0 & A & \gamma^{-2}BB^* & 0 & 0 \\ 0 & 0 & 0 & C^*C & -C^*C & -A^* & -C^* & 0 \\ \hline 0 & 0 & 0 & -C & 0 & 0 & I & 0 \\ 0 & -\gamma^{-2}B^* & 0 & 0 & \gamma^{-2}B^*\Sigma_{21}^* & -\gamma^{-2}B^*\Sigma_{11}^* & 0 & -\gamma^{-2}I \end{array}\right]$$

$$= \left[\begin{array}{cccccc|cc} A & \gamma^{-2}BB^* & 0 & 0 & 0 & 0 & 0 & \gamma^{-2}B \\ -C^*C & -A^* & 0 & 0 & 0 & 0 & 0 & 0 \\ 0 & 0 & -A^* & 0 & 0 & 0 & C^* & -\gamma^{-2}\Sigma_{21}B \\ 0 & 0 & -\gamma^{-2}BB^* & A & 0 & 0 & 0 & \gamma^{-2}\Sigma_{11}B \\ 0 & 0 & 0 & 0 & A & \gamma^{-2}BB^* & 0 & -\gamma^{-2}\Sigma_{11}B \\ 0 & 0 & 0 & 0 & -C^*C & -A^* & 0 & -\gamma^{-2}\Sigma_{21}B \\ \hline 0 & 0 & 0 & -C & 0 & 0 & I & 0 \\ 0 & -\gamma^{-2}B^* & \gamma^{-2}B^*\Sigma_{11}^* & \gamma^{-2}B^*\Sigma_{21}^* & \gamma^{-2}B^*\Sigma_{21}^* & -\gamma^{-2}B^*\Sigma_{11}^* & 0 & -\gamma^{-2}I \end{array}\right]$$

$$= \left[\begin{array}{cc|cc} A & \gamma^{-2}BB^* & 0 & \gamma^{-2}B \\ -C^*C & -A^* & 0 & 0 \\ \hline 0 & 0 & 0 & 0 \\ 0 & -\gamma^{-2}B^* & 0 & 0 \end{array}\right] + \left[\begin{array}{cc|cc} A & \gamma^{-2}BB^* & 0 & -\gamma^{-2}\Sigma_{11}B \\ -C^*C & -A^* & 0 & -\gamma^{-2}\Sigma_{21}B \\ \hline 0 & 0 & 0 & 0 \\ \gamma^{-2}B^*\Sigma_{21}^* & -\gamma^{-2}B^*\Sigma_{11}^* & 0 & 0 \end{array}\right]$$

$$+ \left[\begin{array}{cc|cc} -A^* & 0 & C^* & -\gamma^{-2}\Sigma_{21}B \\ -\gamma^{-2}BB^* & A & 0 & \gamma^{-2}\Sigma_{11}B \\ \hline 0 & -C & I & 0 \\ \gamma^{-2}B^*\Sigma_{11}^* & \gamma^{-2}B^*\Sigma_{21}^* & 0 & -\gamma^{-2}I \end{array}\right]$$

$$= \left[\begin{array}{cc} 0 & 0 \\ 0 & -\gamma^{-2}B \end{array}\right]\left((sI-H)^{-1} - \Sigma^{-1}(sI-H)^{-1}\Sigma\right)\left[\begin{array}{cc} 0 & \gamma^{-2}B \\ 0 & 0 \end{array}\right]$$

$$+ \left[\begin{array}{cc|cc} A & \gamma^{-2}BB^* & 0 & -\gamma^{-2}\Sigma_{11}B \\ 0 & -A^* & C^* & -\gamma^{-2}\Sigma_{21}B \\ \hline C & 0 & I & 0 \\ -\gamma^{-2}B^*\Sigma_{21}^* & \gamma^{-2}B^*\Sigma_{11}^* & 0 & -\gamma^{-2}I \end{array}\right]$$

$$= \left[\begin{array}{cc|cc} A & \gamma^{-2}BB^* & 0 & \gamma^{-2}\Sigma_{11}B \\ 0 & -A^* & -C^* & \gamma^{-2}\Sigma_{21}B \\ \hline -C & 0 & I & 0 \\ \gamma^{-2}B^*\Sigma_{21}^* & -\gamma^{-2}B^*\Sigma_{11}^* & 0 & -\gamma^{-2}I \end{array}\right].$$

The A-matrix of Θ is $\Sigma\begin{bmatrix} A & 0 \\ -C^*C & -A^* \end{bmatrix}\Sigma^{-1}$ and, furthermore, Θ can be obtained, by inverting Θ^{-1} and then applying a similarity transformation Σ, as

$$\Theta = \left[\begin{array}{cc|cc} A & 0 & -\Sigma_{12}^*C^* & B \\ -C^*C & -A^* & \Sigma_{11}^*C^* & 0 \\ \hline -C\Sigma_{11} & -C\Sigma_{12} & I & 0 \\ 0 & B^* & 0 & -\gamma^2 I \end{array}\right].$$

5.7 J-spectral Co-factor of Θ^{-1}

Since there always exist stabilising solutions $L_c \leq 0$ and $L_o \leq 0$ to the two ARE (5.6) and (5.7), respectively, the stable eigenspace of H_c (i.e., the A-matrix of Θ^{-1}) is $\text{Im}\begin{bmatrix} I \\ L_c \end{bmatrix}$ and the antistable eigenspace of H_o is $\text{Im}\begin{bmatrix} L_o \\ I \end{bmatrix}$. The antistable eigenspace of $\Sigma H_o \Sigma^{-1}$ (i.e., the A-matrix of Θ) is

5.7 J-spectral Co-factor of Θ^{-1}

Im $\Sigma \begin{bmatrix} L_o \\ I \end{bmatrix} =$ Im $\begin{bmatrix} \Sigma_{11}L_o + \Sigma_{12} \\ \Sigma_{21}L_o + \Sigma_{22} \end{bmatrix}$. Using similar arguments to those in Chapter 4, Θ^{-1} admits a J-spectral co-factorisation $\Theta^{-1} = W_0^{-1} J W_0^{-\sim}$ if and only if

$$\det \begin{bmatrix} I & \Sigma_{11}L_o + \Sigma_{12} \\ L_c & \Sigma_{21}L_o + \Sigma_{22} \end{bmatrix} = \det \left(\begin{bmatrix} -L_c & I \end{bmatrix} \Sigma \begin{bmatrix} L_o \\ I \end{bmatrix} \right) \neq 0.$$

If this condition holds, then the following matrix is well defined:

$$L_{co} = \begin{bmatrix} I & 0 \end{bmatrix} \Sigma \begin{bmatrix} L_o \\ I \end{bmatrix} \left(\begin{bmatrix} -L_c & I \end{bmatrix} \Sigma \begin{bmatrix} L_o \\ I \end{bmatrix} \right)^{-1}.$$

Using the properties of Σ in Lemma 3.14, it can be shown that $L_{co} = L_{co}^*$ and

$$L_{co} = L_{co}^* = \left(\begin{bmatrix} -L_c & I \end{bmatrix} \Sigma \begin{bmatrix} L_o \\ I \end{bmatrix} \right)^{-*} \begin{bmatrix} L_o & I \end{bmatrix} \Sigma^* \begin{bmatrix} I \\ 0 \end{bmatrix}$$

$$= \left(\begin{bmatrix} L_o & I \end{bmatrix} \Sigma^* \begin{bmatrix} -L_c \\ I \end{bmatrix} \right)^{-1} \begin{bmatrix} L_o & I \end{bmatrix} \Sigma^* \begin{bmatrix} I \\ 0 \end{bmatrix}$$

$$= -\left(\begin{bmatrix} I & -L_o \end{bmatrix} \Sigma^{-1} \begin{bmatrix} I \\ L_c \end{bmatrix} \right)^{-1} \begin{bmatrix} I & -L_o \end{bmatrix} \Sigma^{-1} \begin{bmatrix} 0 \\ I \end{bmatrix}.$$

Instead of directly using Lemma 4.1 and/or Theorem 4.2, two steps will be used to derive the J-spectral co-factor of Θ^{-1}.

(i) Stabilisation: After applying a similarity transformation $\begin{bmatrix} I & 0 \\ L_c & I \end{bmatrix}$,

$$\Theta^{-1} = \left[\begin{array}{cc|cc} A + \gamma^{-2}BB^*L_c & \gamma^{-2}BB^* & 0 & \gamma^{-2}\Sigma_{11}B \\ 0 & -A^* - L_c\gamma^{-2}BB^* & -C^* & \gamma^{-2}(\Sigma_{21} - L_c\Sigma_{11})B \\ \hline -C & 0 & I & 0 \\ \gamma^{-2}B^*(\Sigma_{21}^* - \Sigma_{11}^*L_c) & -\gamma^{-2}B^*\Sigma_{11}^* & 0 & -\gamma^{-2}I \end{array} \right].$$

(ii) Factorisation: After applying another similarity transformation $\begin{bmatrix} I & L_{co} \\ 0 & I \end{bmatrix}$,

$$\Theta^{-1} = \left[\begin{array}{cc|cc} A + \gamma^{-2}BB^*L_c & \theta_{12} & L_{co}C^* & \gamma^{-2}(\Sigma_{11} - L_{co}\Sigma_{21} + L_{co}L_c\Sigma_{11})B \\ 0 & -A^* - \gamma^{-2}L_cBB^* & -C^* & \gamma^{-2}(\Sigma_{21} - L_c\Sigma_{11})B \\ \hline -C & -CL_{co} & I & 0 \\ \gamma^{-2}B^*(\Sigma_{21}^* - \Sigma_{11}^*L_c) & -\gamma^{-2}B^* \begin{bmatrix} 0 & I \end{bmatrix} \Sigma^{-1} \begin{bmatrix} L_{co} \\ I + L_cL_{co} \end{bmatrix} & 0 & -\gamma^{-2}I \end{array} \right].$$

where the element θ_{12} is

$$\theta_{12} = \gamma^{-2}BB^* + L_{co}A^* + L_{co}L_c\gamma^{-2}BB^* + AL_{co} + \gamma^{-2}BB^*L_cL_{co}.$$

It can be verified that the following identity holds:

$$\theta_{12} \equiv -L_{co}C^*CL_{co} + (L_{co}\begin{bmatrix} -L_c & I \end{bmatrix} \Sigma \begin{bmatrix} I \\ 0 \end{bmatrix} - \Sigma_{11})\gamma^{-2}BB^*(L_{co}\begin{bmatrix} -L_c & I \end{bmatrix} \Sigma \begin{bmatrix} I \\ 0 \end{bmatrix} - \Sigma_{11})^*,$$

which is equivalent to

$$\begin{bmatrix} I + L_{co}L_c & -L_{co} \end{bmatrix} \Sigma H_o \Sigma^{-1} \begin{bmatrix} L_{co} \\ I + L_cL_{co} \end{bmatrix} \equiv 0,$$

because of the ARE (5.7) and the following two equalities:

$$\begin{bmatrix} L_{co} \\ \overline{I + L_c L_{co}} \end{bmatrix} = \begin{bmatrix} I & 0 \\ 0 & I \end{bmatrix} \Sigma \begin{bmatrix} L_o \\ I \end{bmatrix} ([-L_c \ I] \Sigma \begin{bmatrix} L_o \\ I \end{bmatrix})^{-1},$$

$$[I + L_{co}L_c \ -L_{co}] = ([I \ -L_o] \Sigma^{-1} \begin{bmatrix} I \\ L_c \end{bmatrix})^{-1} [I \ -L_o] \Sigma^{-1} \begin{bmatrix} I & 0 \\ 0 & I \end{bmatrix}. \quad (5.18)$$

It is a remarkable property that the matrices $\begin{bmatrix} I & 0 \\ 0 & I \end{bmatrix}$ and $\begin{bmatrix} I & 0 \\ 0 & I \end{bmatrix}$ are the identity matrix. As a result, Θ^{-1} can be factorised as

$$\Theta^{-1} = W_0^{-1} J W_0^{-\sim}$$

with W_0^{-1} given by

$$W_0^{-1} = \left[\begin{array}{c|cc} A + \gamma^{-2}BB^*L_c & L_{co}C^* & (L_{co}\,[-L_c\ I]\,\Sigma\begin{bmatrix}I\\0\end{bmatrix} - \Sigma_{11})B\gamma^{-1} \\ \hline -C & I & 0 \\ \gamma^{-2}B^*(\Sigma_{21}^* - \Sigma_{11}^*L_c) & 0 & \gamma^{-1}I \end{array} \right]. $$

(5.19)

As shown below, W_0 is also stable and hence W_0^{-1} is bistable. In other words, it is indeed a J-spectral co-factor of Θ^{-1}. The A-matrix of W_0 is

$$A + \gamma^{-2}BB^*L_c + L_{co}C^*C + (L_{co}\,[-L_c\ I]\,\Sigma\begin{bmatrix}I\\0\end{bmatrix} - \Sigma_{11})\gamma^{-2}BB^*(\Sigma_{11}^*L_c - \Sigma_{21}^*)$$

$$= A + \gamma^{-2}BB^*L_c + L_{co}C^*C + [-I - L_{co}L_c\ L_{co}]\,\Sigma\begin{bmatrix}I\\0\end{bmatrix}\gamma^{-2}BB^*\,[0\ I]\,\Sigma^{-1}\begin{bmatrix}I\\L_c\end{bmatrix}$$

$$= A + \gamma^{-2}BB^*L_c + L_{co}C^*C - ([-L_c\ I]\,\Sigma\begin{bmatrix}L_o\\I\end{bmatrix})^{-*} \cdot$$

$$[I - L_o]\begin{bmatrix}0 & \gamma^{-2}BB^*\\0 & 0\end{bmatrix}\Sigma^{-1}\begin{bmatrix}I\\L_c\end{bmatrix}$$

$$= A + \gamma^{-2}BB^*L_c - ([L_o\ I]\,\Sigma^*\begin{bmatrix}-L_c\\I\end{bmatrix})^{-1} \cdot$$

$$[I - L_o]\,\Sigma^{-1}\left\{ \begin{bmatrix}0\\C^*C\end{bmatrix} + \Sigma\begin{bmatrix}0 & \gamma^{-2}BB^*\\0 & 0\end{bmatrix}\Sigma^{-1}\begin{bmatrix}I\\L_c\end{bmatrix} \right\}$$

$$= A + \gamma^{-2}BB^*L_c - ([I - L_o]\,\Sigma^{-1}\begin{bmatrix}I\\L_c\end{bmatrix})^{-1}[I - L_o]\,\Sigma^{-1}(H_c - \Sigma H_o \Sigma^{-1})\begin{bmatrix}I\\L_c\end{bmatrix}$$

$$= ([I - L_o]\,\Sigma^{-1}\begin{bmatrix}I\\L_c\end{bmatrix})^{-1}[I - L_o]\,\Sigma^{-1} \cdot$$

$$\left\{ \begin{bmatrix}I\\L_c\end{bmatrix}(A + \gamma^{-2}BB^*L_c) - (H_c - \Sigma H_o \Sigma^{-1})\begin{bmatrix}I\\L_c\end{bmatrix} \right\}$$

$$= ([I - L_o]\,\Sigma^{-1}\begin{bmatrix}I\\L_c\end{bmatrix})^{-1}[I - L_o]\,\Sigma^{-1}\left\{ \Sigma H_o \Sigma^{-1}\begin{bmatrix}I\\L_c\end{bmatrix} \right\}$$

$$\sim [I - L_o]\,H_o \Sigma^{-1}\begin{bmatrix}I\\L_c\end{bmatrix}([I - L_o]\,\Sigma^{-1}\begin{bmatrix}I\\L_c\end{bmatrix})^{-1}$$

$$= [\,I\ -L_o\,]\,H_o\left\{\begin{bmatrix}I&0\\0&I\end{bmatrix}+\begin{bmatrix}L_o\\I\end{bmatrix}[\,0\ -I\,]\right\}\Sigma^{-1}\begin{bmatrix}I\\L_c\end{bmatrix}([\,I\ -L_o\,]\,\Sigma^{-1}\begin{bmatrix}I\\L_c\end{bmatrix})^{-1}$$

$$= [\,I\ -L_o\,]\,H_o\begin{bmatrix}I&-L_o\\0&0\end{bmatrix}\Sigma^{-1}\begin{bmatrix}I\\L_c\end{bmatrix}([\,I\ -L_o\,]\,\Sigma^{-1}\begin{bmatrix}I\\L_c\end{bmatrix})^{-1}$$

$$= [\,I\ -L_o\,]\,H_o\begin{bmatrix}I\\0\end{bmatrix},$$

where "$\sim$" means similar to. The addition of ARE (5.7) was used just after the similarity transformation.

Due to (5.18), W_0^{-1} can be further simplified to

$$W_0^{-1} = \left[\begin{array}{c|cc} A+\gamma^{-2}BB^*L_c & L_{co}C^* - [\,I+L_{co}L_c\ -L_{co}\,]\,\Sigma\begin{bmatrix}I\\0\end{bmatrix} & B\gamma^{-1} \\ \hline -C & I & 0 \\ \gamma^{-2}B^*(\Sigma_{21}^* - \Sigma_{11}^*L_c) & 0 & \gamma^{-1}I \end{array}\right]$$

$$= \left[\begin{array}{c|cc} A+\gamma^{-2}BB^*L_c & \hat{\Sigma}_{22}^{-*}(\Sigma_{12}^* + L_o\Sigma_{11}^*)C^* & -\hat{\Sigma}_{22}^{-*}B\gamma^{-1} \\ \hline -C & I & 0 \\ \gamma^{-2}B^*(\Sigma_{21}^* - \Sigma_{11}^*L_c) & 0 & \gamma^{-1}I \end{array}\right],$$

where

$$\hat{\Sigma}_{22} = [\,-L_c\ I\,]\,\Sigma\begin{bmatrix}L_o\\I\end{bmatrix}, \qquad \hat{\Sigma}_{22}^* = [\,I\ -L_o\,]\,\Sigma^{-1}\begin{bmatrix}I\\L_c\end{bmatrix}.$$

5.8 A Numerical Example

A numerical example[3] is given here. Consider

$$G_\beta(s) = -\frac{1}{s-a}$$

with a minimal realization

$$G_\beta = \left[\begin{array}{c|c} a & 1 \\ \hline -1 & 0 \end{array}\right],$$

i.e., $A = a$, $B = 1$ and $C = 1$. According to the assumptions, $a \neq 0$. This gives

$$H_c = \begin{bmatrix} a & \gamma^{-2} \\ 0 & -a \end{bmatrix}, \quad H_o = \begin{bmatrix} a & 0 \\ -1 & -a \end{bmatrix} \quad \text{and} \quad H = \begin{bmatrix} a & \gamma^{-2} \\ -1 & -a \end{bmatrix}.$$

The AREs (5.6) and (5.7) can be re-written as

$$L_c^2\gamma^{-2} + 2aL_c = 0 \quad \text{and} \quad L_o^2 + 2aL_o = 0.$$

---
[3] Portions reprinted, with permission, from [168]. ©IEEE.

When $a < 0$, i.e., G_β is stable, the stabilising solutions are

$$L_c = 0 \quad \text{and} \quad L_o = 0.$$

When $a > 0$, i.e., G_β is unstable, the stabilising solutions are[4]

$$L_c = -2a\gamma^2 \quad \text{and} \quad L_o = -2a.$$

It is easy to find that the eigenvalues of H are $\lambda_{1,2} = \pm\lambda$ with $\lambda = \sqrt{a^2 - \gamma^{-2}}$. Note that λ may be an imaginary number. Assume that $\gamma \neq 1/|a|$ temporarily. With two similarity transformations with

$$S_1 = \begin{bmatrix} 1 & -a+\lambda \\ 0 & 1 \end{bmatrix} \quad \text{and} \quad S_2 = \begin{bmatrix} 1 & 0 \\ -\frac{1}{2\lambda} & 1 \end{bmatrix},$$

H can be transformed into $\begin{bmatrix} \lambda & 0 \\ 0 & -\lambda \end{bmatrix}$, i.e.,

$$S_2^{-1} S_1^{-1} H S_1 S_2 = \begin{bmatrix} \lambda & 0 \\ 0 & -\lambda \end{bmatrix}.$$

Hence, the Σ-matrix is

$$\Sigma = e^{Hh} = S_1 S_2 \begin{bmatrix} e^{\lambda h} & 0 \\ 0 & e^{-\lambda h} \end{bmatrix} S_2^{-1} S_1^{-1}$$

$$= S_1 \begin{bmatrix} e^{\lambda h} & 0 \\ -\frac{1}{2\lambda}(e^{\lambda h} - e^{-\lambda h}) & e^{-\lambda h} \end{bmatrix} S_1^{-1}$$

$$= \begin{bmatrix} e^{\lambda h} + \frac{a-\lambda}{2\lambda}(e^{\lambda h} - e^{-\lambda h}) & \frac{a^2-\lambda^2}{2\lambda}(e^{\lambda h} - e^{-\lambda h}) \\ -\frac{1}{2\lambda}(e^{\lambda h} - e^{-\lambda h}) & e^{-\lambda h} - \frac{a-\lambda}{2\lambda}(e^{\lambda h} - e^{-\lambda h}) \end{bmatrix}.$$

When $\gamma = 1/|a|$, i.e., $\lambda = 0$, the above Σ still holds if the limit for $\lambda \to 0$ is taken on the right-hand side, which gives

$$\Sigma|_{\lambda=0} = \begin{bmatrix} 1+ah & a^2 h \\ -h & 1-ah \end{bmatrix}.$$

In the sequel, it is assumed that Σ is defined as above for $\lambda = 0$.

5.8.1 The Stable Case ($a < 0$)

In this case, G_β is stable and $L_c = L_o = 0$. Hence, $\hat{\Sigma}_{22}$ defined in (5.8) is

$$\hat{\Sigma}_{22} = \Sigma_{22} = e^{-\lambda h} - \frac{a-\lambda}{2\lambda}(e^{\lambda h} - e^{-\lambda h}).$$

---

[4] In this case, $A + \gamma^{-2} B B^* L_c = a - 2a = -a$ and $A + L_o C^* C = a - 2a = -a$ are stable.

When $\gamma \geq 1/|a|$, the number λ is positive and hence the eigenvalues of H are real. It is easy to see that $\hat{\Sigma}_{22}$ is always positive (nonsingular).[5] According to Theorem 5.1,
$$\gamma_{opt} < 1/|a|.$$
This reflects the right half of the condition (5.5).

When $0 < \gamma < 1/|a|$, the number $\lambda = \omega i$, where $\omega = \sqrt{\gamma^{-2} - a^2}$, is imaginary and hence the eigenvalues of H are imaginaries. However, $\hat{\Sigma}_{22}$ is still a real number because
$$\hat{\Sigma}_{22} = e^{-\omega h i} - \frac{a - \omega i}{2\omega i}(e^{\omega h i} - e^{-\omega h i})$$
$$= \cos(\omega h) - \frac{a}{\omega}\sin(\omega h).$$

Substitute $\omega = \sqrt{\gamma^{-2} - a^2}$ into this, then
$$\hat{\Sigma}_{22} = \cos(\frac{ah}{a\gamma}\sqrt{1 - a^2\gamma^2}) - \frac{a\gamma}{\sqrt{1 - a^2\gamma^2}}\sin(\frac{ah}{a\gamma}\sqrt{1 - a^2\gamma^2}).$$

This can be regarded as a surface, as shown in Figure 5.2, with respect to the normalised delay ah and the normalised performance index $a\gamma$. This surface crosses the plane $\hat{\Sigma}_{22} = 0$ many times, as can be seen from the contours of $\hat{\Sigma}_{22} = 0$ shown in Figure 5.3. The top curve in Figure 5.3 characterises the normalised optimal performance index $a\gamma_{opt}$ with respect to the normalised delay ah. On this curve, $\hat{\Sigma}_{22}$ becomes singular the first time when γ decreases from $+\infty$ (or actually, $\|G_\beta\|_{L_\infty}$) to 0. Since $\|\Gamma_{G_\beta}\| = 0$ and $\|G_\beta\|_{L_\infty} = 1/|a|$, the optimal value γ_{opt} satisfies $0 \leq \gamma_{opt} \leq 1/|a|$, i.e., $-1 \leq a\gamma_{opt} \leq 0$. This coincides with the curve $a\gamma_{opt}$ shown in Figure 5.3.

The (suboptimal) compensator for $\gamma > \gamma_{opt}$ is discussed below.

According to (5.10), (5.12), (5.13), (5.15), and (5.16),
$$Z(s) = -\pi_h(\begin{bmatrix} a & \gamma^{-2} & 0 \\ -1 & -a & 1 \\ 0 & \gamma^{-2} & 0 \end{bmatrix}) = \gamma^{-2}\frac{\gamma^{-2}\Sigma_{21}^* + (e^{-sh} - \Sigma_{11}^*)(s - a)}{s^2 + \gamma^{-2} - a^2}. \qquad (5.20)$$

The locus of the hidden poles of $Z(s)$, which are the eigenvalues of H, is shown in Figure 5.4. When $\gamma \geq 1/|a|$, $Z(s)$ has two hidden real poles symmetric to the $j\omega$-axis. When $0 < \gamma < 1/|a|$, $Z(s)$ has a pair of hidden imaginary poles. In either case, the implementation of Z needs care; see Part II.

The W^{-1} given in (5.11) is well defined for $\gamma > \gamma_{opt}$ as
$$W^{-1} = \begin{bmatrix} a & \Sigma_{22}^{-*}\Sigma_{12}^* & -\Sigma_{22}^{-*} \\ -1 & 1 & 0 \\ \gamma^{-2}\Sigma_{21}^* & 0 & 1 \end{bmatrix}$$

---
[5] $\hat{\Sigma}_{22} = 1 - ah$ when $\gamma = 1/|a| = -1/a$.

5 The Delay-type Nehari Problem

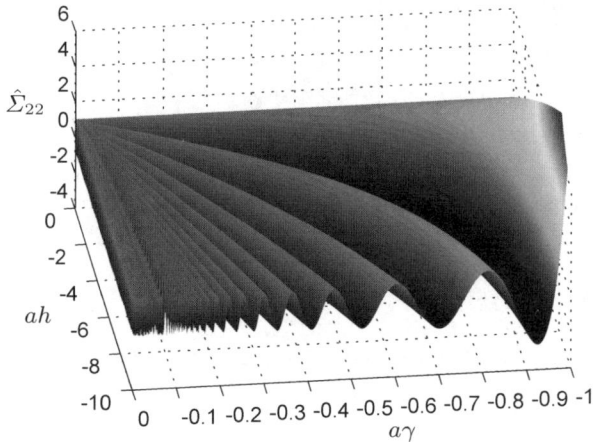

Figure 5.2. Surface of $\hat{\Sigma}_{22}$ with respect to ah and $a\gamma$ $(a < 0)$

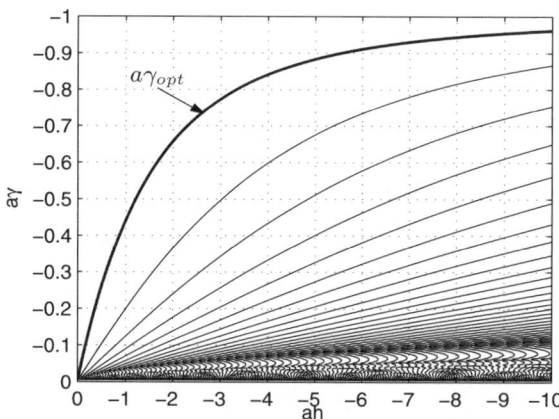

Figure 5.3. Contour $\hat{\Sigma}_{22} = 0$ on the ah-$a\gamma$ plane $(a < 0)$

and $\Pi_{22} = \begin{bmatrix} Z & I \end{bmatrix} W^{-1} \begin{bmatrix} 0 \\ I \end{bmatrix}$ is

$$\Pi_{22} = \begin{bmatrix} Z & 1 \end{bmatrix} \left[\begin{array}{c|c} a & -\Sigma_{22}^{-*} \\ \hline -1 & 0 \\ \gamma^{-2}\Sigma_{21}^* & 1 \end{array}\right] = 1 + \frac{\Sigma_{22}^{-*}}{s-a}(Z - \gamma^{-2}\Sigma_{21}^*)$$

$$= 1 + \gamma^{-2}\Sigma_{22}^{-*}\frac{e^{-sh} - \Sigma_{11}^* - \Sigma_{21}^*(s+a)}{s^2 + \gamma^{-2} - a^2}.$$

It is easy to see that Π_{22} is stable. As a matter of fact, as required, Π_{22} is bistable for $\gamma > \gamma_{opt}$. The Nyquist plots of Π_{22} for different values of $a\gamma$ are

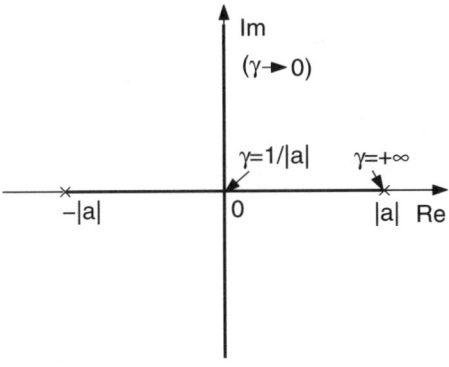

Figure 5.4. Locus of the hidden poles of Z

shown in Figure 5.5 for $ah = -1$. The optimal value γ_{opt} is between $-0.44/a$ and $-0.45/a$. This corresponds to the transition from Figure 5.5(c) to Figure 5.5(d), in which the number of encirclements changes according to the change of the bi-stability of Π_{22}: the Nyquist plot encircles the origin when $\gamma < \gamma_{opt}$ and hence Π_{22} is not bistable but the Nyquist plot does not encircle the origin when $\gamma > \gamma_{opt}$ and hence Π_{22} is bistable.

5.8.2 The Unstable Case ($a > 0$)

In this case, G_β is unstable and $L_c = -2a\gamma^2$, $L_o = -2a$. Hence, $\hat{\Sigma}_{22}$ defined in (5.8) is

$$\hat{\Sigma}_{22} = e^{-\lambda h} - 4a^2\gamma^2 e^{\lambda h} + \frac{-2a\gamma^2\lambda^2 - 2a^3\gamma^2 + 4\lambda a^2\gamma^2 + a + \lambda}{2\lambda}(e^{\lambda h} - e^{-\lambda h})$$

$$= \frac{-2a\gamma^2\lambda^2 - 2a^3\gamma^2 + a + \lambda - 4\lambda a^2\gamma^2}{2\lambda}e^{\lambda h}$$

$$- \frac{-2a\gamma^2\lambda^2 - 2a^3\gamma^2 + a - \lambda + 4\lambda a^2\gamma^2}{2\lambda}e^{-\lambda h}$$

$$= \frac{-2a\gamma^2\lambda^2 - 2a^3\gamma^2 + a}{2\lambda}(e^{\lambda h} - e^{-\lambda h}) + \frac{e^{\lambda h} + e^{-\lambda h}}{2}(1 - 4a^2\gamma^2)$$

$$= \frac{-4a^3\gamma^2 + 3a}{2\lambda}(e^{\lambda h} - e^{-\lambda h}) + \frac{e^{\lambda h} + e^{-\lambda h}}{2}(1 - 4a^2\gamma^2). \quad (5.21)$$

$\hat{\Sigma}_{22}$ can be re-arranged as

$$\hat{\Sigma}_{22} = -\frac{4a(a^2\gamma^2 - 1) + a}{2\lambda}(e^{\lambda h} - e^{-\lambda h}) - \frac{4(a^2\gamma^2 - 1) + 3}{2}(e^{\lambda h} + e^{-\lambda h}).$$

Hence, $\hat{\Sigma}_{22}$ is always negative (nonsingular)[6] when $\gamma \geq 1/|a|$, noting that a and λ are positive. According to Theorem 5.1, the optimal performance index

---
[6] $\hat{\Sigma}_{22} = -ah - 3$ when $\gamma = 1/|a| = 1/a$.

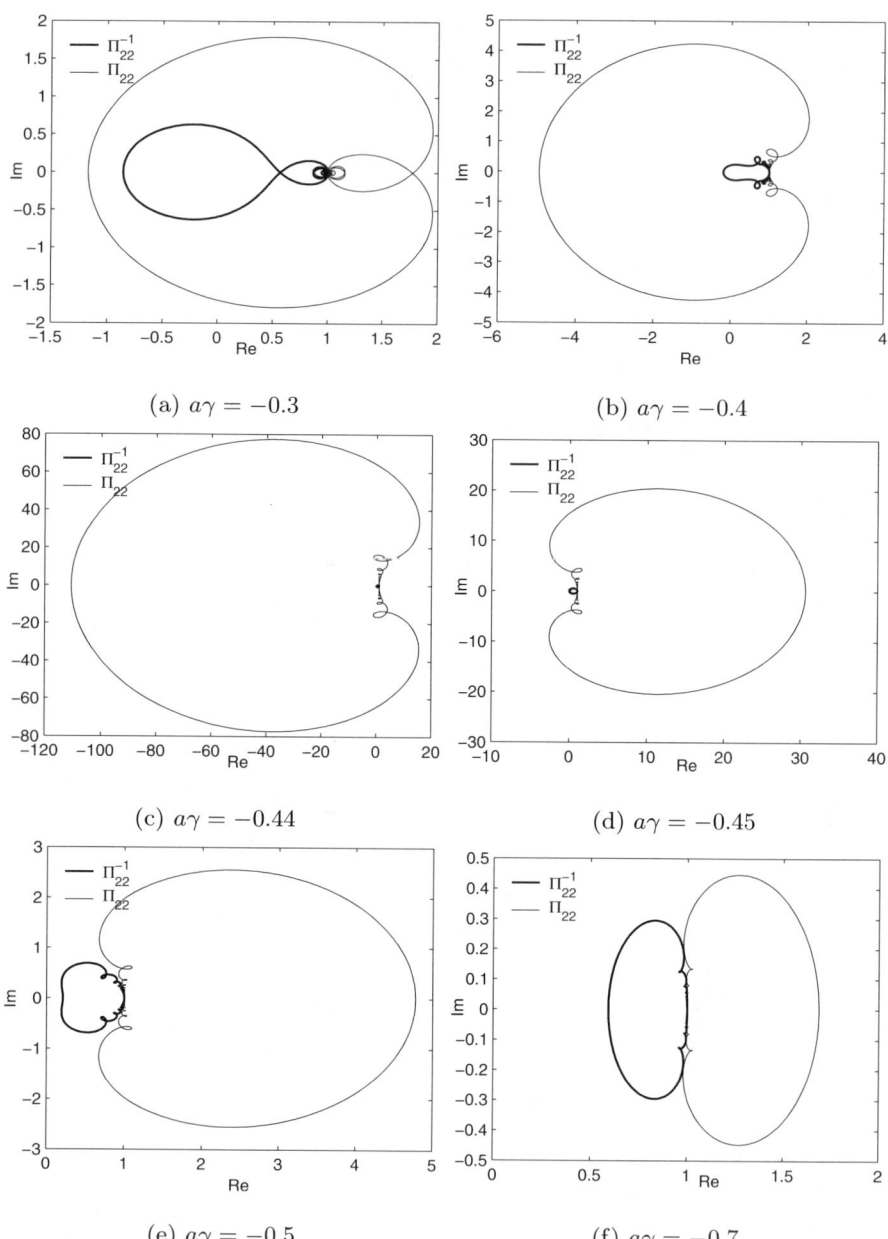

Figure 5.5. Nyquist plots of Π_{22} ($a < 0$ and $ah = -1$)

is less than $1/|a|$, i.e.,
$$\gamma_{opt} < 1/|a|.$$
This actually reflects the right half of the condition (5.5).

When $0 < \gamma < 1/|a|$, the eigenvalues $\pm\lambda$ of H are on the $j\omega$-axis. Substitute $\lambda = \omega i$ with $\omega = \sqrt{\gamma^{-2} - a^2}$ into (5.21), then
$$\hat{\Sigma}_{22} = (1 - 4a^2\gamma^2)\cos(\omega h) + (3 - 4a^2\gamma^2)\frac{a}{\omega}\sin(\omega h).$$

Similarly, with the substitution $\omega = \sqrt{\gamma^{-2} - a^2} = \gamma^{-1}\sqrt{1 - a^2\gamma^2}$, then
$$\hat{\Sigma}_{22} = (1 - 4a^2\gamma^2)\cos(\frac{ah}{a\gamma}\sqrt{1 - a^2\gamma^2}) + \frac{a\gamma(3 - 4a^2\gamma^2)}{\sqrt{1 - a^2\gamma^2}}\sin(\frac{ah}{a\gamma}\sqrt{1 - a^2\gamma^2}).$$

This surface is shown in Figure 5.6(b) and the contours of $\hat{\Sigma}_{22} = 0$ on the ah-$a\gamma$ plane are shown in Figure 5.7(b). The top curve in Figure 5.7(b) characterises the normalised optimal performance index $a\gamma_{opt}$ with respect to the normalised delay ah. On this curve, $\hat{\Sigma}_{22}$ becomes singular the first time when γ decreases from $+\infty$ to 0.

Since $I - L_c L_o = 1 - 4a^2\gamma^2$, then $\|\Gamma_{G_\beta}\| = \frac{1}{2a}$. As a result, the optimal value γ_{opt} satisfies
$$\frac{1}{2a} \leq \gamma_{opt} \leq \frac{1}{a},$$
i.e., $0.5 \leq a\gamma_{opt} \leq 1$. This coincides with the curve shown in Figure 5.7(b).

The (suboptimal) compensator for $\gamma > \gamma_{opt}$ is discussed below.

In this case, the FIR block Z remains the same as in (5.20) and the form is not affected because of the sign of a. However, W^{-1} is changed to

$$W^{-1} = \left[\begin{array}{c|cc} -a & \hat{\Sigma}_{22}^{-*}(\Sigma_{12}^* - 2a\Sigma_{11}^*) & -\hat{\Sigma}_{22}^{-*} \\ -1 & 1 & 0 \\ \gamma^{-2}\Sigma_{21}^* + 2a\Sigma_{11}^* & 0 & 1 \end{array}\right]$$

and $\Pi_{22} = \begin{bmatrix} Z & I \end{bmatrix} W^{-1} \begin{bmatrix} 0 \\ I \end{bmatrix}$ is

$$\Pi_{22} = \begin{bmatrix} Z & 1 \end{bmatrix} \left[\begin{array}{c|cc} -a & -\hat{\Sigma}_{22}^{-*} \\ -1 & 0 \\ \gamma^{-2}\Sigma_{21}^* + 2a\Sigma_{11}^* & 1 \end{array}\right]$$

$$= 1 + \frac{\hat{\Sigma}_{22}^{-*}}{s+a}(Z - \gamma^{-2}\Sigma_{21}^* - 2a\Sigma_{11}^*)$$

$$= 1 + \gamma^{-2}\hat{\Sigma}_{22}^{-*} \cdot \frac{\frac{s-a}{s+a}e^{-sh} - (s-a)\Sigma_{21}^* + (2a^2\gamma^2 - 1 - 2a\gamma^2 s)\Sigma_{11}^*}{s^2 + \gamma^{-2} - a^2}$$

$$= 1 + \gamma^{-2}\hat{\Sigma}_{22}^{-*} \cdot \frac{\frac{s-a}{s+a}e^{-sh} - (s-a)(\Sigma_{21}^* + 2a\gamma^2\Sigma_{11}^*) - \Sigma_{11}^*}{s^2 + \gamma^{-2} - a^2}.$$

106   5 The Delay-type Nehari Problem

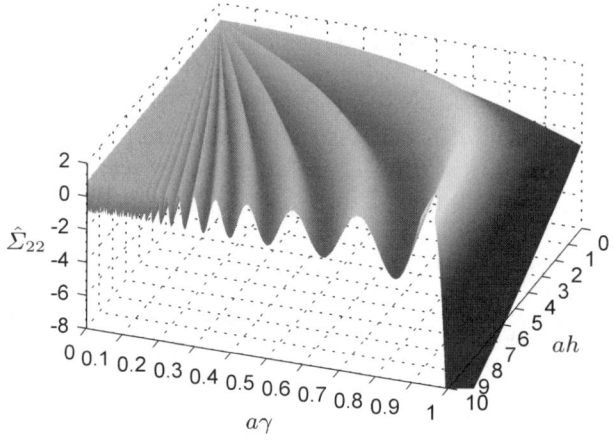

Figure 5.6. Surface of $\hat{\Sigma}_{22}$ with respect to ah and $a\gamma$ ($a > 0$)

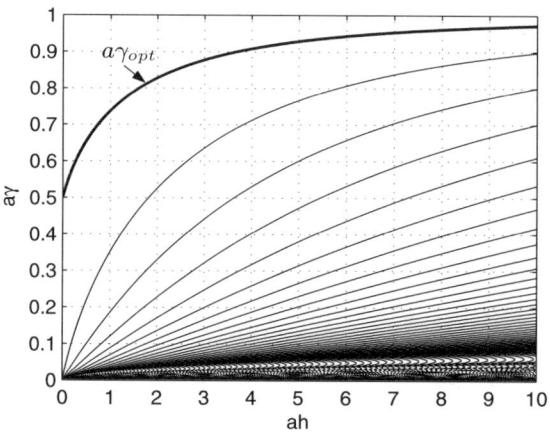

Figure 5.7. Contour $\hat{\Sigma}_{22} = 0$ on the ah-$a\gamma$ plane ($a > 0$)

Apparently, Π_{22} is stable and invertible. As a matter of fact, as required, Π_{22} is bistable for $\gamma > \gamma_{opt}$. The Nyquist plots of Π_{22} for different values of $a\gamma$ are shown in Figure 5.8 for $ah = 1$. The optimal value γ_{opt} is between $0.73/a$ and $0.74/a$. This corresponds to the transition from Figure 5.8(c) to Figure 5.8(d), in which the number of encirclements changes accordingly to the change of the bi-stability of Π_{22}.

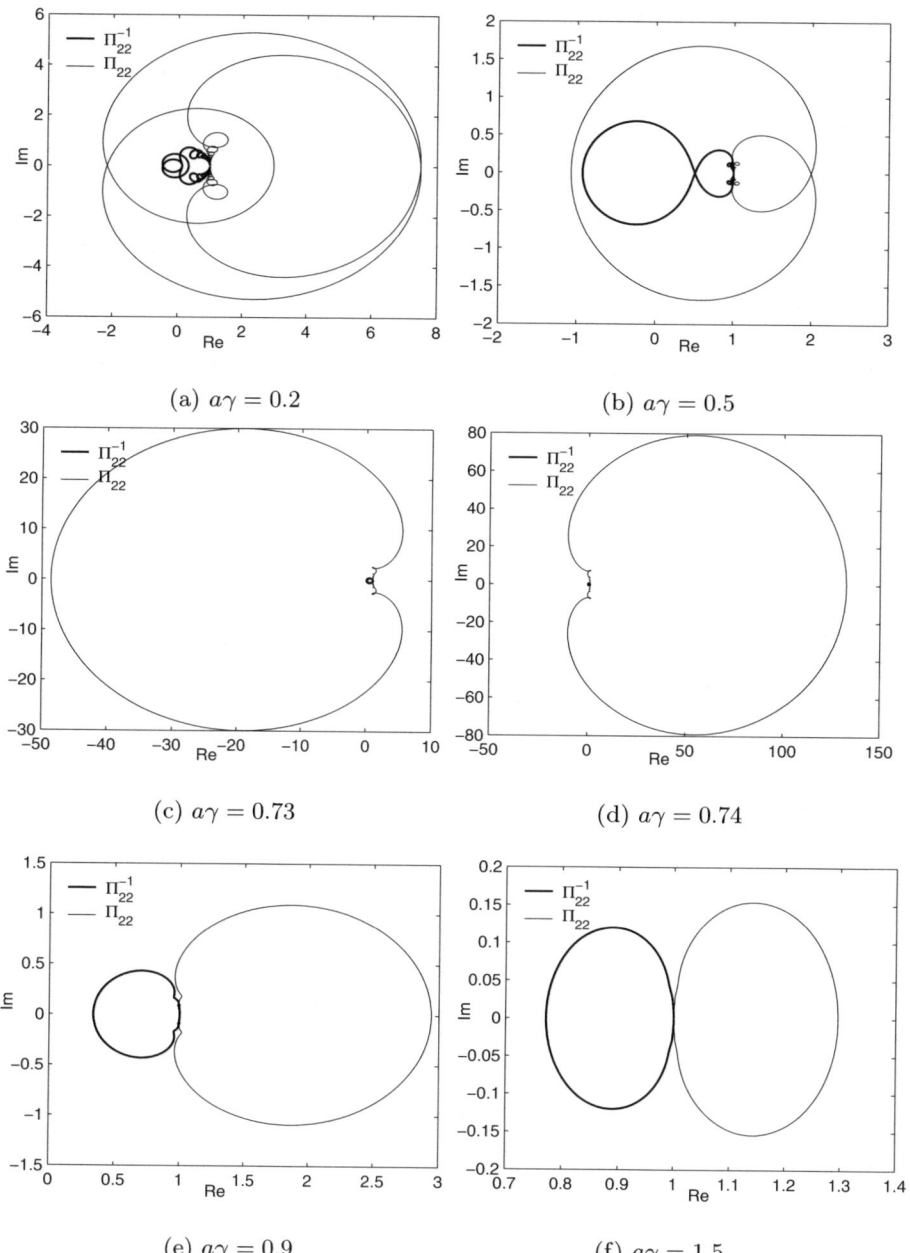

Figure 5.8. Nyquist plots of Π_{22} ($a > 0$ and $ah = 1$)

5.9 Summary and Notes

This chapter derives the solution to the delay-type Nehari problem. This generalises the result for the stable case [182] to the unstable case. This is also an extension of the conventional (delay-free) Nehari problem to the case with a delay. The solvability condition is formulated in terms of the nonsingularity of a matrix.

The solution to a more general problem (but for the stable case) where the delay is replaced by an inner function $m(s)$ was conjectured in [182], which was then proved in [66] by using the skew Toeplitz theory. A simple proof in the s-domain was then given in [125]. Another simple and elegant proof was given in [154]. An easier formula for computation was given in [56] where m is a pseudo-rational transfer function and it was then extended to the mixed sensitivity problem in [55].

6
An Extended Nehari Problem

In this chapter[1], a different type of Nehari problem with a delay is considered. Here, instead of the requirement on the stability of the compensator K, the stability of the closed-loop transfer matrix is required. Hence, the norm involved in this chapter is the H^∞-norm rather than the L^∞-norm. As will be seen in the next chapter, the solution to this problem is vital for solving the standard H^∞ problem with a delay. While the solvability condition of this problem is well known, the parameterisation of all the suboptimal compensators is not trivial.

6.1 Problem Statement

The extended Nehari problem with a delay (ENP$_h$) is formulated as follows.

> Given a $\gamma > 0$ and a strictly proper but not necessarily stable $G_{\beta 11}$, find a proper (but not necessarily stable) Q_β such that $G_{\beta 11} + e^{-sh} Q_\beta$ is stable and
> $$\left\| G_{\beta 11} + e^{-sh} Q_\beta \right\|_\infty < \gamma. \qquad (6.1)$$

If $Q_\beta(s) \in H^\infty$ is required, then the H^∞-norm in (6.1) becomes the L^∞-norm and the problem becomes the delay-type Nehari problem (NP$_h$) studied in the previous chapter. When $G_{\beta 11}$ is stable, this problem is the same as the NP$_h$ with a stable G_β.

---

[1] Portions reprinted, with permission, from [162]. ©IEEE.

6.2 The Solvability Condition

The ENP_h (6.1) is actually a stabilisation problem. It can be converted to a conventional Nehari problem

$$\|e^{sh}Q_Z + Q_0\|_\infty < \gamma, \tag{6.2}$$

where $Q_0 = \tilde{G}_{\beta 11} + Q_\beta$ is stable and $Q_Z = G_{\beta 11} - e^{-sh}\tilde{G}_{\beta 11} = \mathcal{T}_h\{G_{\beta 11}\}$ is a (stable) FIR block. This was obtained by Mirkin [85] as a constraint imposed on the free parameter of the delay-free H^∞ controllers for processes with a delay. An alternative implicit solution can be found in [87], which seems not straightforward.

It is well-known [36, XXXV.4] that this problem (*i.e.*, the ENP_h) is solvable iff $\gamma > \gamma_h \doteq \|\Gamma_{e^{sh}Q_Z}\|$, where Γ denotes the Hankel operator. Inspecting the transfer matrix $e^{sh}Q_Z$, one can see that γ_h is actually the $L_2[0, h]$-induced norm of $G_{\beta 11}$, *i.e.*, $\gamma_h = \|G_{\beta 11}\|_{L_2[0,h]}$. Various methods [30, 41, 182] have been proposed to compute this norm; see Section 3.6 for more details. A simple representation of γ_h is the maximal γ making Σ_{22} singular [182], *i.e.*,

$$\gamma_h = \max\{\gamma : \det \Sigma_{22} = 0\}, \tag{6.3}$$

which is now known as the Zhou–Khargonekar formula [55].

6.3 Solution

Theorem 6.1. *Given* $G_{\beta 11} = \left[\begin{array}{c|c} A & B \\ \hline -C & 0 \end{array}\right]$, *there exists an admissible solution to the extended Nehari problem with a delay iff* $\gamma > \gamma_h$. *Furthermore, if the above condition holds, then all* Q_β *satisfying (6.1) are parameterised as*

$$Q_\beta = \mathcal{H}_r(\begin{bmatrix} I & 0 \\ Z_1 & I \end{bmatrix} W^{-1}, \; Q), \tag{6.4}$$

where

$$Z_1 = -\pi_h\left\{\left[\begin{array}{cc|c} A & \gamma^{-2}BB^* & 0 \\ -C^*C & -A^* & C^* \\ \hline 0 & \gamma^{-2}B^* & 0 \end{array}\right]\right\}, \tag{6.5}$$

$$W^{-1} = \left[\begin{array}{c|cc} A & \Sigma_{12}\Sigma_{22}^{-1}C^* & -\Sigma_{22}^{-*}B \\ \hline -C & I & 0 \\ \gamma^{-2}B^*\Sigma_{21}^* & 0 & I \end{array}\right], \tag{6.6}$$

and $Q(s)$ *is a stable free parameter with* $\|Q(s)\|_\infty < \gamma$.

Remark 6.1. When $G_{\beta 11}$ is stable, the ENP_h is the same as the NP_h with a stable G_β. Indeed, the solution of the ENP_h is the same as that of NP_h in this case given in Corollary 5.1 (Subsection 5.5.1) for a stable G_β.

6.4 Proof

The first part of Theorem 6.1 has been shown in Section 6.2. The only thing left to be proved is to derive the parameterisation of Q_β under the condition $\gamma > \gamma_h$.

6.4.1 Rationalisation by Z_1

The ENP_h problem (6.1) can be associated with the following system in chain-scattering representation:

$$\begin{bmatrix} z_1 \\ w_1 \end{bmatrix} = G(s) \begin{bmatrix} u_2 \\ z_2 \end{bmatrix} \doteq \begin{bmatrix} e^{-sh}I & G_{\beta 11} \\ 0 & I \end{bmatrix} \begin{bmatrix} u_2 \\ z_2 \end{bmatrix} \text{ and } u_2 = Q_\beta z_2 \qquad (6.7)$$

and hence can be re-written as $\|\mathcal{H}_r(G, Q_\beta)\|_\infty < \gamma$, as shown in Figure 6.1. It is well known [39, 58, 80] that the H^∞ control problem is closely related to the matrix $G^\sim J_\gamma G$. Moreover, the H^∞ problem (6.1) is solved once $G(s)$ is completed to be a J-lossless transfer matrix.

Meinsma and Zwart [80] studied a 2-block problem, where G is stable. The result cannot be directly used to solve the extended Nehari problem with a delay because G is not necessarily stable and, more importantly, neither is Q_β. However, the ideas are borrowed.

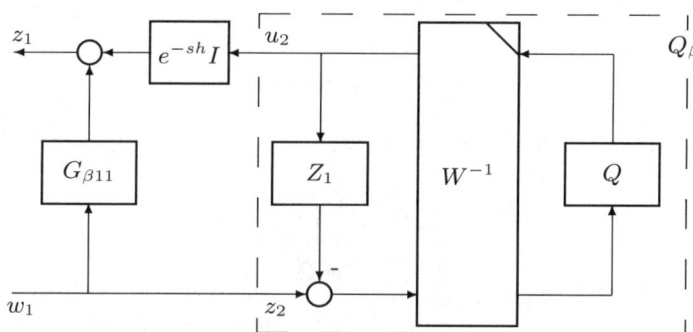

Figure 6.1. Solving the ENP_h

By introducing the FIR block

$$Z_1(s) = F_0(s) + F_1(s)e^{-sh} \doteq -\pi_h\{\mathcal{F}_u(\begin{bmatrix} G_{\beta 11} & I \\ I & 0 \end{bmatrix}, \gamma^{-2}G_{\widetilde{\beta}11})\}, \qquad (6.8)$$

the following matrix:

$$\Theta \doteq \begin{bmatrix} I & Z_1^\sim \\ 0 & I \end{bmatrix} G^\sim J_\gamma G \begin{bmatrix} I & 0 \\ Z_1 & I \end{bmatrix}$$

becomes rational (bearing in mind that $\begin{bmatrix} I & 0 \\ Z_1 & I \end{bmatrix}$ is a unimodular matrix). Using elementary transformations, Θ^{-1} can be obtained (see Section 5.6) as

$$\Theta^{-1} = \left[\begin{array}{cc|cc} A & \gamma^{-2}BB^* & 0 & \gamma^{-2}\Sigma_{11}B \\ 0 & -A^* & -C^* & \gamma^{-2}\Sigma_{21}B \\ \hline -C & 0 & I & 0 \\ \gamma^{-2}B^*\Sigma_{21}^* & -\gamma^{-2}B^*\Sigma_{11}^* & 0 & -\gamma^{-2}I \end{array}\right].$$

6.4.2 Completing the J-losslessness

Θ^{-1} can be further factorised as

$$\Theta^{-1} = W^{-1} \cdot J_\gamma^{-1} \cdot W^{-\sim},$$

where W^{-1} is given in (6.6), without the need to solve *any* Riccati equation[2] but using the Σ-related identities given in Lemma 3.14. The following transfer matrix:

$$M = \begin{bmatrix} M_{11} & M_{12} \\ M_{21} & M_{22} \end{bmatrix} \doteq G \begin{bmatrix} I & 0 \\ Z_1 & I \end{bmatrix} W^{-1},$$

of which the realization is given in Section 6.5, is stable and J-unitary. Furthermore, it is J-lossless because its (2,2)-block M_{22} can be shown to be bistable using similar arguments to those in [80, p.284: the second part of Condition 2 in the Proof of Theorem 5.3]. It can now be concluded, according to Theorem 6.2 in [80, p.281], that

$$G_{\beta 11} + e^{-sh}Q_\beta = \mathcal{H}_r(M, Q)$$

is stable and $\|G_{\beta 11} + e^{-sh}Q_\beta\|_\infty < \gamma$ for

$$Q_\beta = \mathcal{H}_r(\begin{bmatrix} I & 0 \\ Z_1 & I \end{bmatrix} W^{-1}, Q) \tag{6.9}$$

with any stable $Q(s)$ having $\|Q(s)\|_\infty < \gamma$. This condition is also necessary, as demonstrated below by applying the arguments in [78]. By construction,

$$\|G_{\beta 11} + e^{-sh}Q_\beta\|_{L_\infty} < \gamma \quad \text{if and only if} \quad \|Q(s)\|_{L_\infty} < \gamma$$

for

$$Q = \mathcal{H}_r(W \begin{bmatrix} I & 0 \\ -Z_1 & I \end{bmatrix}, Q_\beta).$$

Since

$$\lim_{\operatorname{Re} s \geq 0, |s| \to \infty} W(s) \begin{bmatrix} I & 0 \\ -Z_1(s) & I \end{bmatrix} = \begin{bmatrix} I & 0 \\ 0 & I \end{bmatrix},$$

---

[2] This is a notable characteristic. Due to this, $G_{\beta 11}$ may have $j\omega$-axis poles and/or zeros.

it is true that
$$Q(\infty) = \mathcal{H}_r(\begin{bmatrix} I & 0 \\ 0 & I \end{bmatrix}, Q_\beta(\infty)) = Q_\beta(\infty).$$

This implies that $Q(s)$ given above is proper iff $Q_\beta(s)$ is proper, yet the set of proper operators in L^∞ is in fact H^∞ [137] (see also [21, A6.26.c, A6.27]). So if Q_β solves ENP_h, which means Q_β is proper, then necessarily $\|Q(s)\|_\infty < \gamma$.

The only thing left to be proved is the state-space realization of Z_1, which can be found as given in (6.5) using the formula in Subsection 3.3.1.

6.5 Realization of M

Since
$$M = \begin{bmatrix} M_{11} & M_{12} \\ M_{21} & M_{22} \end{bmatrix} = G \begin{bmatrix} I & 0 \\ Z_1 & I \end{bmatrix} W^{-1}$$
$$= \begin{bmatrix} e^{-sh}W_{11} + G_{\beta 11}(Z_1 W_{11} + W_{21}) & e^{-sh}W_{12} + G_{\beta 11}(Z_1 W_{12} + W_{22}) \\ Z_1 W_{11} + W_{21} & Z_1 W_{12} + W_{22} \end{bmatrix}$$

where W^{-1} as given in (6.6) is re-written as $\begin{bmatrix} W_{11} & W_{12} \\ W_{21} & W_{22} \end{bmatrix}$ according to the original partition, M can be obtained by elements as follows:

(i) M_{21}: Since $Z_1(s) = F_0(s) + F_1(s)e^{-sh}$,
$$M_{21} = \begin{bmatrix} F_0 & I \end{bmatrix} \begin{bmatrix} W_{11} \\ W_{21} \end{bmatrix} + F_1 W_{11} e^{-sh},$$

where
$$\begin{bmatrix} F_0 & I \end{bmatrix} \begin{bmatrix} W_{11} \\ W_{21} \end{bmatrix} = \left[\begin{array}{cc|cc} A & \gamma^{-2}BB^* & 0 & 0 \\ -C^*C & -A^* & C^* & 0 \\ \hline \gamma^{-2}B^*\Sigma_{21}^* & -\gamma^{-2}B^*\Sigma_{11}^* & 0 & I \end{array}\right] \left[\begin{array}{c|c} A & \Sigma_{12}\Sigma_{22}^{-1}C^* \\ -C & I \\ \hline \gamma^{-2}B^*\Sigma_{21}^* & 0 \end{array}\right]$$
$$= \left[\begin{array}{ccc|c} A & \gamma^{-2}BB^* & 0 & 0 \\ -C^*C & -A^* & -C^*C & C^* \\ 0 & 0 & A & \Sigma_{12}\Sigma_{22}^{-1}C^* \\ \hline \gamma^{-2}B^*\Sigma_{21}^* & -\gamma^{-2}B^*\Sigma_{11}^* & \gamma^{-2}B^*\Sigma_{21}^* & 0 \end{array}\right]$$
$$= \left[\begin{array}{ccc|c} A & \gamma^{-2}BB^* & 0 & 0 \\ 0 & -A^* & -C^*C & C^* \\ 0 & \gamma^{-2}BB^* & A & \Sigma_{12}\Sigma_{22}^{-1}C^* \\ \hline 0 & -\gamma^{-2}B^*\Sigma_{11}^* & \gamma^{-2}B^*\Sigma_{21}^* & 0 \end{array}\right]$$
$$= \left[\begin{array}{cc|c} A & \gamma^{-2}BB^* & 0 \\ -C^*C & -A^* & \Sigma_{22}^{-1}C^* \\ \hline 0 & -\gamma^{-2}B^* & 0 \end{array}\right],$$

and

$$F_1 W_{11} = \left[\begin{array}{cc|c} A & \gamma^{-2}BB^* & \Sigma_{12}\Sigma_{22}^{-1}C^* \\ -C^*C & -A^* & C^* \\ \hline 0 & \gamma^{-2}B^* & 0 \end{array}\right].$$

According to the definition of the truncation operator τ_h,

$$M_{21} = \tau_h \left\{ \left[\begin{array}{cc|c} A & \gamma^{-2}BB^* & 0 \\ -C^*C & -A^* & \Sigma_{22}^{-1}C^* \\ \hline 0 & -\gamma^{-2}B^* & 0 \end{array}\right] \right\}.$$

(ii) $M_{11} = e^{-sh}W_{11} + G_{\beta 11} M_{21}$, in which the delay-free term is

$$\left[\begin{array}{ccc|c} A & \gamma^{-2}BB^* & 0 & 0 \\ -C^*C & -A^* & 0 & \Sigma_{22}^{-1}C^* \\ 0 & -\gamma^{-2}BB^* & A & 0 \\ \hline 0 & 0 & -C & 0 \end{array}\right] = \left[\begin{array}{cc|c} A & \gamma^{-2}BB^* & 0 \\ -C^*C & -A^* & \Sigma_{22}^{-1}C^* \\ \hline C & 0 & 0 \end{array}\right]$$

and the delay term is

$$W_{11} + G_{\beta 11} \cdot \left[\begin{array}{cc|c} A & \gamma^{-2}BB^* & \Sigma_{12}\Sigma_{22}^{-1}C^* \\ -C^*C & -A^* & C^* \\ \hline 0 & \gamma^{-2}B^* & 0 \end{array}\right]$$

$$= I + \left[\begin{array}{c|c} A & \Sigma_{12}\Sigma_{22}^{-1}C^* \\ \hline -C & 0 \end{array}\right] + \left[\begin{array}{ccc|c} A & \gamma^{-2}BB^* & 0 & \Sigma_{12}\Sigma_{22}^{-1}C^* \\ -C^*C & -A^* & 0 & C^* \\ 0 & \gamma^{-2}BB^* & A & 0 \\ \hline 0 & 0 & -C & 0 \end{array}\right]$$

$$= I + \left[\begin{array}{cc|c} A & \gamma^{-2}BB^* & \Sigma_{12}\Sigma_{22}^{-1}C^* \\ -C^*C & -A^* & C^* \\ \hline -C & 0 & 0 \end{array}\right].$$

According to the definition of the truncation operator τ_h,

$$M_{11} = e^{-sh}I + \tau_h \left\{ \left[\begin{array}{cc|c} A & \gamma^{-2}BB^* & 0 \\ -C^*C & -A^* & \Sigma_{22}^{-1}C^* \\ \hline C & 0 & 0 \end{array}\right] \right\}.$$

(iii) M_{22}: Since $Z_1(s) = F_0(s) + F_1(s)e^{-sh}$, then

$$M_{22} = I + \begin{bmatrix} F_0 & I \end{bmatrix} \begin{bmatrix} W_{12} \\ W_{22} - I \end{bmatrix} + F_1 W_{12} e^{-sh},$$

where

$$\begin{bmatrix} F_0 & I \end{bmatrix} \begin{bmatrix} W_{12} \\ W_{22} - I \end{bmatrix} = \left[\begin{array}{cc|cc} A & \gamma^{-2}BB^* & 0 & 0 \\ -C^*C & -A^* & C^* & 0 \\ \hline \gamma^{-2}B^*\Sigma_{21}^* & -\gamma^{-2}B^*\Sigma_{11}^* & 0 & I \end{array}\right] \left[\begin{array}{c|c} A & -\Sigma_{22}^{-*}B \\ -C & 0 \\ \hline \gamma^{-2}B^*\Sigma_{21}^* & 0 \end{array}\right]$$

$$= \begin{bmatrix} A & \gamma^{-2}BB^* & 0 & 0 \\ -C^*C & -A^* & -C^*C & 0 \\ 0 & 0 & A & -\Sigma_{22}^{-*}B \\ \hline \gamma^{-2}B^*\Sigma_{21}^* & -\gamma^{-2}B^*\Sigma_{11}^* & \gamma^{-2}B^*\Sigma_{21}^* & 0 \end{bmatrix}$$

$$= \begin{bmatrix} A & \gamma^{-2}BB^* & 0 & -\Sigma_{22}^{-*}B \\ -C^*C & -A^* & 0 & 0 \\ 0 & 0 & A & -\Sigma_{22}^{-*}B \\ \hline \gamma^{-2}B^*\Sigma_{21}^* & -\gamma^{-2}B^*\Sigma_{11}^* & 0 & 0 \end{bmatrix}$$

$$= \begin{bmatrix} A & \gamma^{-2}BB^* & -\Sigma_{22}^{-*}B \\ -C^*C & -A^* & 0 \\ \hline \gamma^{-2}B^*\Sigma_{21}^* & -\gamma^{-2}B^*\Sigma_{11}^* & 0 \end{bmatrix}$$

$$= \begin{bmatrix} A & \gamma^{-2}BB^* & -B \\ -C^*C & -A^* & \Sigma_{22}^{-1}\Sigma_{21}B \\ \hline 0 & -\gamma^{-2}B^* & 0 \end{bmatrix},$$

using a similarity transform with Σ at the last step, and

$$F_1 W_{12} = \begin{bmatrix} A & \gamma^{-2}BB^* & -\Sigma_{22}^{-*}B \\ -C^*C & -A^* & 0 \\ \hline 0 & \gamma^{-2}B^* & 0 \end{bmatrix}.$$

According to the definition of the truncation operator τ_h,

$$M_{22} = I + \tau_h \left\{ \begin{bmatrix} A & \gamma^{-2}BB^* & -B \\ -C^*C & -A^* & \Sigma_{22}^{-1}\Sigma_{21}B \\ \hline 0 & -\gamma^{-2}B^* & 0 \end{bmatrix} \right\}.$$

(iv) $M_{12} = e^{-sh} W_{12} + G_{\beta 11} M_{22}$, in which the delay-free term is

$$G_{\beta 11} + G_{\beta 11} \cdot \begin{bmatrix} A & \gamma^{-2}BB^* & -B \\ -C^*C & -A^* & \Sigma_{22}^{-1}\Sigma_{21}B \\ \hline 0 & -\gamma^{-2}B^* & 0 \end{bmatrix}$$

$$= \begin{bmatrix} A & \gamma^{-2}BB^* & 0 & -B \\ -C^*C & -A^* & 0 & \Sigma_{22}^{-1}\Sigma_{21}B \\ 0 & -\gamma^{-2}BB^* & A & B \\ \hline 0 & 0 & -C & 0 \end{bmatrix}$$

$$= \begin{bmatrix} A & \gamma^{-2}BB^* & -B \\ -C^*C & -A^* & \Sigma_{22}^{-1}\Sigma_{21}B \\ \hline C & 0 & 0 \end{bmatrix}$$

and the delay term is

$$W_{12} + G_{\beta 11} \cdot \begin{bmatrix} A & \gamma^{-2}BB^* & -\Sigma_{22}^{-*}B \\ -C^*C & -A^* & 0 \\ \hline 0 & \gamma^{-2}B^* & 0 \end{bmatrix}$$

$$= \begin{bmatrix} A & -\Sigma_{22}^{-*}B \\ \hline -C & 0 \end{bmatrix} + \begin{bmatrix} A & \gamma^{-2}BB^* & 0 & -\Sigma_{22}^{-*}B \\ -C^*C & -A^* & 0 & 0 \\ 0 & \gamma^{-2}BB^* & A & 0 \\ \hline 0 & 0 & -C & 0 \end{bmatrix}$$

$$= \begin{bmatrix} A & \gamma^{-2}BB^* & -\Sigma_{22}^{-*}B \\ -C^*C & -A^* & 0 \\ \hline -C & 0 & 0 \end{bmatrix}.$$

According to the definition of the truncation operator τ_h,

$$M_{12} = \tau_h \left\{ \begin{bmatrix} A & \gamma^{-2}BB^* & -B \\ -C^*C & -A^* & \Sigma_{22}^{-1}\Sigma_{21}B \\ \hline C & 0 & 0 \end{bmatrix} \right\}.$$

Combining the above results, M is then obtained as

$$M = \begin{bmatrix} e^{-sh}I & 0 \\ 0 & I \end{bmatrix} + \tau_h \left\{ \begin{bmatrix} A & \gamma^{-2}BB^* & 0 & -B \\ -C^*C & -A^* & \Sigma_{22}^{-1}C^* & \Sigma_{22}^{-1}\Sigma_{21}B \\ \hline C & 0 & 0 & 0 \\ 0 & -\gamma^{-2}B^* & 0 & 0 \end{bmatrix} \right\}.$$

6.6 Summary

The extended Nehari-problem is solved in this chapter. This is a very similar problem to the delay-type Nehari problem considered in the previous chapter. The difference is that, instead of requiring the stability of the compensator K, the stability of the closed-loop transfer matrix is needed. The solvability condition is known, but the parameterisation of all the suboptimal compensators is not trivial. This result will be used in the next chapter to solve the standard H^∞ problem with a delay.

7

The Standard H^∞ Problem

In this chapter,[1] the standard H^∞ control problem for processes with a single delay is considered. A frequency-domain approach is proposed to split the problem to a standard delay-free H^∞ problem and a one-block problem. The one-block problem is then further reduced to an extended Nehari problem. Hence, for a given bound on the H^∞-norm of the closed-loop transfer function, there exist proper stabilising controllers that achieve this bound if and only if both the corresponding delay-free H^∞ problem and the extended Nehari problem with a delay (or the one-block problem) are all solvable. Applying the results obtained in the previous chapter, the solvability conditions of the standard H^∞ control problem with a delay are formulated in terms of the existence of solutions to two delay-independent algebraic Riccati equations and a nonsingularity property of a delay-dependent matrix. All suboptimal controllers solving the problems are, respectively, parameterised as structures incorporating a modified Smith predictor.

7.1 Introduction

The H^∞ control of processes with delay(s) has been an active research area since the mid 1980s. Early frequency response methods treated such systems within the framework of general infinite-dimensional system theory [30, 31, 115, 138, 139]. This resulted in rather complicated solutions, for which the structure is not transparent. Some problem-oriented approaches [13, 26, 59, 100] were motivated to exploit the structure of systems and simpler structures incorporating finite-impulse-response (FIR) blocks were proposed in [111, 117, 118] for implementation. Although considerable progress has been made in this direction, most of the existing solutions still lack the transparency of the classical predictor-type controller. For example, Nagpal and Ravi [100] obtained remarkably elegant solvability conditions for the general H^∞ control

---

[1] Portions reprinted, with permission, from [162]. ©IEEE.

problem; however, the resulting controllers are extremely complicated. The recent work of Tadmor [134] has made the solution much simpler, but still not very transparent. One of the Riccati equations of [134] depends on the delay as well as on the solution of another differential Riccati equation. A notable exception is the recent work of Meinsma and Zwart [80], who derived the solution of the 2-block mixed sensitivity problem using a Smith-predictor-type controller. This approach was extended to the standard H^∞ problem [79] by reducing the four-block problem to a two-block problem. However, the solution is based on several intermediate model transformations and, as a result, the final formulae are rather involved and the Riccati equation still depends on the delay. Moreover, there is no clear relationship between the solutions of the H^∞ problems for the system with a delay and for its delay-free counterpart. It is not clear how the delay affects the achievable cost or what is the rationale behind the prediction block, especially in the four-block case; see [92] and references therein.

Mirkin [85, 87] treated the delay element not as a part of the generalised plant but as a causality constraint imposed upon the controller and then extracted the controller from the solution of the delay-free counterpart. Using this idea, the four-block problem was reduced to a one-block problem and the relationship with the delay-free counterpart becomes clear. In this approach, it is necessary to find the constraint imposed on the free parameter in the solution to the delay-free problem. This is difficult and offers few hints for the case with multiple delays. Moreover, parameterising the transfer functions to satisfy the desired constraint is not easy.

This chapter presents a more natural and more intuitive approach to solve the problem in the frequency domain. The four-block problem is reduced to a delay-free four-block problem and a one-block problem with a delay by inserting a unimodular matrix and its inverse. According to the chain-scattering theory [58], this transformation does not change the problem (in fact, this idea can be generalised to the multiple-delay case or even some other cases having a constrained controller). The one-block problem with a delay is then reduced to the extended Nehari problem with a delay by inserting a unimodular FIR block. Using the solution to the extended Nehari problem obtained in the previous chapter, the solutions to the one-block problem and to the standard problem are recovered. This also offers a proof for a different (but equivalent) result given in [85, Lemma 2]. An alternative proof can be found in [87].

The rest of this chapter is organised as follows. The problems are stated in Section 7.2. The main problem is reduced to a delay-free problem and an extended Nehari problem with a delay in Section 7.3. The solutions to the problems are given in Section 7.4. Proof of the solution to the main problem (and the one-block problem) is given in Section 7.5.

7.2 Problem Statements

The main problem considered in this chapter is the standard H^∞ control problem for systems with a single delay. This problem is solved via reduction to a one-block problem with a delay and, then, to an extended Nehari problem with a delay.

7.2.1 The Standard H^∞ Problem (SP_h)

The standard H^∞ control problem for systems with a single I/O delay (SP_h) is formulated as follows.

Given a $\gamma > 0$ and the general control setup for processes with a single I/O delay as shown in Figure 7.1, find a proper controller K such that the closed-loop system is internally stable and
$$\left\|\mathcal{F}_l(P,\ Ke^{-sh})\right\|_\infty < \gamma. \tag{7.1}$$

When $h = 0$, the problem becomes the common standard H^∞ control problem (SP_0).

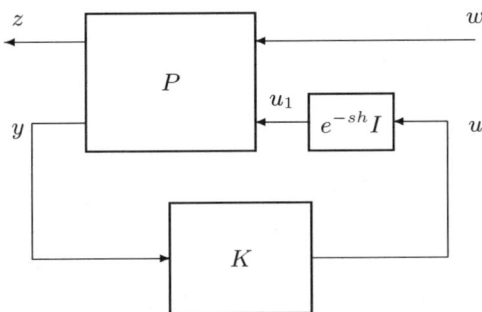

Figure 7.1. General setup of control systems with a single I/O delay

7.2.2 The One-block Problem (OP_h)

The one-block problem for systems with a single delay (OP_h) is formulated as follows.

Given a $\gamma > 0$ and a not necessarily stable G_β with $G_\beta(\infty) = \begin{bmatrix} 0 & I \\ I & 0 \end{bmatrix}$, find a proper (but not necessarily stable) $K(s)$ such that

$$\|\mathcal{F}_l(G_\beta, Ke^{-sh})\|_\infty < \gamma. \tag{7.2}$$

This is actually a special case of the general one-block problem, where the (1, 2) and (2, 1) blocks of $G_\beta(\infty)$ are identity matrices. In order to simplify the exposition, another condition that the (1, 2) and (2, 1) blocks of G_β have a stable inverse will be assumed in this chapter. This is sufficient to derive the solution to the main problem.

7.3 Reduction of the Standard Problem (SP$_h$)

7.3.1 The Standard Delay-free H^∞ Problem (SP$_0$)

Assume that the plant in Figure 7.1 is realized as

$$P(s) = \begin{bmatrix} P_{11}(s) & P_{12}(s) \\ P_{21}(s) & P_{22}(s) \end{bmatrix} = \left[\begin{array}{c|cc} A_p & B_{p1} & B_{p2} \\ \hline C_{p1} & 0 & D_{p12} \\ C_{p2} & D_{p21} & 0 \end{array} \right]$$

and the following standard assumptions hold:

(A1) (A_p, B_{p2}) is stabilisable and (C_{p2}, A_p) is detectable;

(A2) $\begin{bmatrix} A_p - j\omega I & B_{p2} \\ C_{p1} & D_{p12} \end{bmatrix}$ and $\begin{bmatrix} A_p - j\omega I & B_{p1} \\ C_{p2} & D_{p21} \end{bmatrix}$ have full column rank and full row rank, respectively, $\forall \omega \in \mathbb{R}$;

(A3) $D^*_{p12} D_{p12} = I$ and $D_{p21} D^*_{p21} = I$.

Assumption (A3) is made to simplify the exposition although, in fact, only the nonsingularity of the matrices $D^*_{p12} D_{p12}$ and $D_{p21} D^*_{p21}$ is required [40].

When $h = 0$, the problem is reduced to the common standard H^∞ control problem, for which the well-known results are given in [25, 39]. The following two Hamiltonian matrices are involved in the solution:

$$H_0 = \begin{bmatrix} A_p & \gamma^{-2} B_{p1} B^*_{p1} \\ -C^*_{p1} C_{p1} & -A^*_p \end{bmatrix} - \begin{bmatrix} B_{p2} \\ -C^*_{p1} D_{p12} \end{bmatrix} \begin{bmatrix} D^*_{p12} C_{p1} & B^*_{p2} \end{bmatrix},$$

$$J_0 = \begin{bmatrix} A^*_p & \gamma^{-2} C^*_{p1} C_{p1} \\ -B_{p1} B^*_{p1} & -A_p \end{bmatrix} - \begin{bmatrix} C^*_{p2} \\ -B_{p1} D^*_{p21} \end{bmatrix} \begin{bmatrix} D_{p21} B^*_{p1} & C_{p2} \end{bmatrix}.$$

Lemma 7.1. [39] *Assume that the conditions (A1–A3) are satisfied, then there exists an admissible controller K, denoted K_0, such that $\|\mathcal{F}_l(P, K_0)\|_\infty < \gamma$ (i.e., the delay-free case when $h = 0$ in Figure 7.1) iff the following three conditions hold:*

(i) $H_0 \in \text{dom}(\text{Ric})$ and $X = \text{Ric}(H_0) \geq 0$;

(ii) $J_0 \in dom(Ric)$ and $Y = Ric(J_0) \geq 0$;
(iii) $\rho(XY) < \gamma^2$.

Moreover, when these conditions hold, all admissible controllers such that $\|\mathcal{F}_l(P, K_0)\|_\infty < \gamma$ are parameterised as

$$K_0 = \mathcal{H}_r(G_\alpha, Q_\alpha) \qquad (7.3)$$

where

$$G_\alpha = \left[\begin{array}{c|cc} A_\alpha - B_1 C_2 & B_2 & B_1 \\ \hline C_1 & I & 0 \\ -C_2 & 0 & I \end{array}\right]$$

with

$$A_\alpha = A_p + LC_{p2} + \gamma^{-2} Y C_{p1}^* C_{p1} + \left(B_{p2} + \gamma^{-2} Y C_{p1}^* D_{p12}\right) F\Psi,$$

$$B_1 = -L, \qquad B_2 = B_{p2} + \gamma^{-2} Y C_{p1}^* D_{p12},$$

$$C_1 = F\Psi, \qquad C_2 = -\left(C_{p2} + \gamma^{-2} D_{p21} B_{p1}^* X\right)\Psi,$$

$$F = -(B_{p2}^* X + D_{p12}^* C_{p1}), \qquad L = -(Y C_{p2}^* + B_{p1} D_{p21}^*),$$

$$\Psi = (I - \gamma^{-2} YX)^{-1},$$

and $Q_\alpha(s) \in H_\infty$ is a free parameter satisfying

$$\|Q_\alpha(s)\|_\infty < \gamma. \qquad (7.4)$$

7.3.2 Reducing SP$_h$ to OP$_h$

The general setup for control systems with a single I/O delay shown in Figure 7.1 can be equivalently depicted in chain-scattering representation as shown in Figure 7.2(a). Since G_α and $\mathcal{C}_r(G_\beta) \doteq G_\alpha^{-1}$ are all bistable [58, p214], inserting G_α and its inverse $\mathcal{C}_r(G_\beta)$ between the process and the controller, as shown in Figure 7.2(b), does not change the solvability condition of the original problem according to Lemma 3.2. The original four-block H^∞ control problem is then split into a delay-free problem and a one-block problem as given in (7.2), where

$$G_\beta = \begin{bmatrix} G_{\beta 11} & G_{\beta 12} \\ G_{\beta 21} & G_{\beta 22} \end{bmatrix} = \left[\begin{array}{c|cc} A & B_1 & B_2 \\ \hline -C_1 & 0 & I \\ -C_2 & I & 0 \end{array}\right]$$

with $A \doteq A_\alpha - B_1 C_2 - B_2 C_1$. Although A_α and/or A may be unstable but $A + B_2 C_1 = A_\alpha - B_1 C_2$ and $A + B_1 C_2 = A_\alpha - B_2 C_1$ are always stable [58].

In the sequel, because of the slight notation change in the realization of $G_{\beta 11}$ ($B \to B_1$ and $C \to C_1$), H is defined in a slightly different way as

$$H = \begin{bmatrix} A & \gamma^{-2} B_1 B_1^* \\ -C_1^* C_1 & -A^* \end{bmatrix}.$$

7 The Standard H^∞ Problem

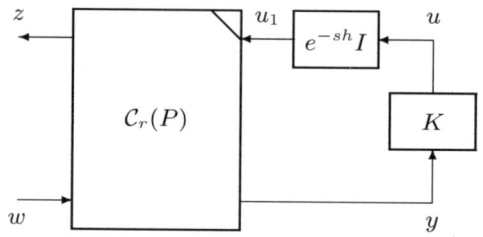

(a) SP_h in chain-scattering representation

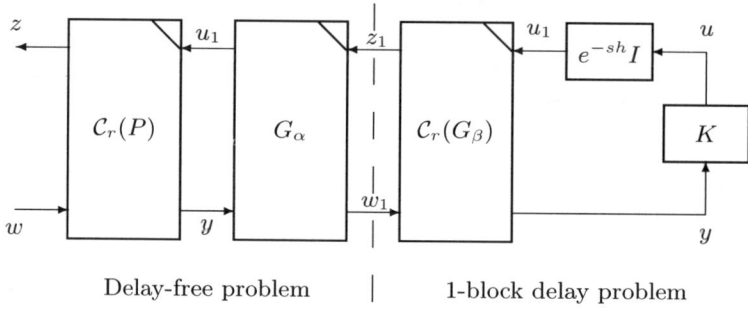

(b) decomposition of SP_h

Figure 7.2. Reduction of SP_h to SP_0 and a one-block delay problem

Note: It is assumed that P_{21} is invertible here, so that $\mathcal{C}_r(P)$ exists. This assumption does not affect the results and can be removed because G_α is not affected by this condition and the block $\mathcal{C}_r(P)$ can be re-formulated in the input–output representation as usual. A tagged *right*-upper corner indicates that the matrix in the block, e.g., G_α, is a right CST and hence there is no confusion.

Correspondingly, the definitions of Σ and γ_h are, respectively,

$$\Sigma = \begin{bmatrix} \Sigma_{11} & \Sigma_{12} \\ \Sigma_{21} & \Sigma_{22} \end{bmatrix} \doteq e^{Hh}, \tag{7.5}$$

and

$$\gamma_h = \max\{\gamma : \det \Sigma_{22} = 0\}. \tag{7.6}$$

7.3.3 Reducing OP_h to ENP_h

The transfer function involved in OP_h is

$$\mathcal{F}_l(G_\beta, Ke^{-sh}) = G_{\beta 11} + e^{-sh}G_{\beta 12}K\left(I - G_{\beta 22}Ke^{-sh}\right)^{-1}G_{\beta 21},$$

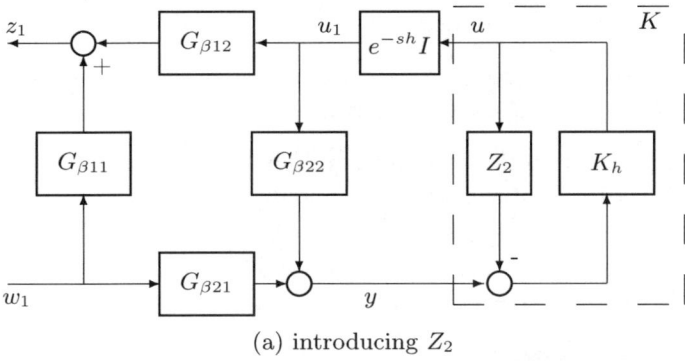

(a) introducing Z_2

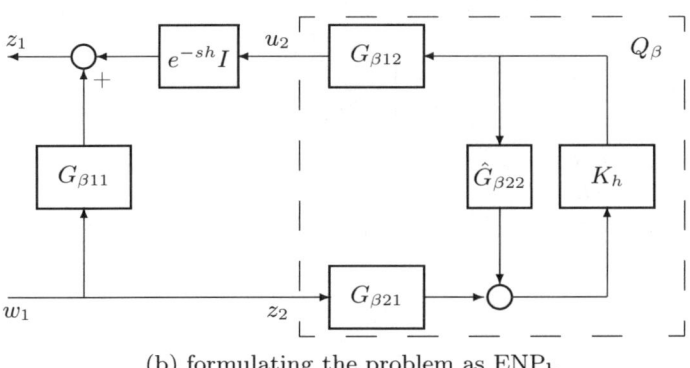

(b) formulating the problem as ENP_h

Figure 7.3. Reducing the one-block problem to ENP_h

which is depicted in Figure 7.3(a). It can be further simplified by introducing a modified Smith predictor-type controller

$$K \doteq K_h(I + Z_2 K_h)^{-1} = \mathcal{H}_r\left(\begin{bmatrix} I & 0 \\ Z_2 & I \end{bmatrix}, K_h\right), \tag{7.7}$$

where $Z_2(s)$ is a (stable) FIR block

$$Z_2 = -\pi_h\{G_{\beta 22}\} \doteq -\hat{G}_{\beta 22} + e^{-sh} G_{\beta 22} \tag{7.8}$$

and $\begin{bmatrix} I & 0 \\ Z_2 & I \end{bmatrix}$ is a unimodular matrix. Hence, this transformation does not change the solvability of the problem, according to Lemma 3.2, and the OP_h given in (7.2) is then converted to the ENP_h given in (5.3), as shown in Figure 7.3(b), where

$$Q_\beta \doteq \mathcal{H}_r\left(\begin{bmatrix} G_{\beta 12} & 0 \\ -G_{\beta 21}^{-1} \hat{G}_{\beta 22}(s) & G_{\beta 21}^{-1} \end{bmatrix}, K_h\right). \tag{7.9}$$

This means that there is a bijection from proper Q_β to proper K_h. Hence, the following lemma is now obtained:

Lemma 7.2. SP_h *is solvable iff its delay-free counterpart* SP_0 *and the corresponding* ENP_h *(or* OP_h*) are all solvable.*

This transformation is not necessary. However, it may offer some useful hints for solving the problem with multiple delays.

7.4 Solutions

7.4.1 Solution to OP_h

Theorem 7.1. *For a given*

$$G_\beta = \left[\begin{array}{c|cc} A & B_1 & B_2 \\ \hline -C_1 & 0 & I \\ C_2 & I & 0 \end{array}\right]$$

with stable $A + B_2C_1$ *and* $A + B_1C_2$*, there exists an admissible solution to* OP_h *iff* $\gamma > \gamma_h$*, where* γ_h *is defined in (7.6).*

Furthermore, if this condition holds, then all K *satisfying (7.2) are parameterised as*

$$K = \mathcal{H}_r(\begin{bmatrix} I & 0 \\ Z & I \end{bmatrix} V^{-1}, Q) \tag{7.10}$$

where

$$Z = -\pi_h\{\left[\begin{array}{cc|c} A & \gamma^{-2}B_1B_1^* & B_2 \\ -C_1^*C_1 & -A^* & C_1^* \\ \hline -C_2 & \gamma^{-2}B_1^* & 0 \end{array}\right]\}, \tag{7.11}$$

$$V^{-1} = \left[\begin{array}{c|cc} A + B_2C_1 & B_2 - \Sigma_{12}\Sigma_{22}^{-1}C_1^* & \Sigma_{22}^{-*}B_1 \\ \hline C_1 & I & 0 \\ -\gamma^{-2}B_1^*\Sigma_{21}^* - C_2\Sigma_{22}^* & 0 & I \end{array}\right], \tag{7.12}$$

and $\|Q(s)\|_\infty < \gamma$ *is a free parameter.* $\Sigma = \begin{bmatrix} \Sigma_{11} & \Sigma_{12} \\ \Sigma_{21} & \Sigma_{22} \end{bmatrix}$ *is defined in (7.5).*

7.4.2 Solution to SP_h

Theorem 7.2. *Assume that the conditions (A1–A3) hold,* SP_h *is solvable iff the following conditions hold:*

(i) $H_0 \in dom(Ric)$ *and* $X = Ric(H_0) \geq 0$;
(ii) $J_0 \in dom(Ric)$ *and* $Y = Ric(J_0) \geq 0$;
(iii) $\rho(XY) < \gamma^2$;
(iv) $\gamma > \gamma_h$.

Furthermore, if these conditions hold, then all K *satisfying (7.1) are given in (7.10) with (7.11) and (7.12).*

Remark 7.1. The solvability conditions of the standard H^∞ control problem for systems with a single delay are those for the delay-free case plus the non-singularity of Σ_{22}. This result is very compact.

Remark 7.2. For delay-free systems (*i.e.*, $h = 0$), $\Sigma \equiv I$. $\Sigma_{22} \equiv I$ is always nonsingular for any $\gamma > 0$, condition (iv) disappears and V^{-1} becomes G_α. Hence, condition (iv) can be regarded as the additional cost of the delay.

Remark 7.3. V is bistable because $\Sigma_{22}^{-*}(A + B_1 C_2)\Sigma_{22}^* \sim A + B_1 C_2$, which is also stable [58].

Remark 7.4. This theorem indicates that $C_r(P)\begin{bmatrix} I & 0 \\ Z & I \end{bmatrix} V^{-1}$ is J-lossless and, hence, its lower-right block is bistable.

7.5 Proof

The first part of Theorems 7.1 and 7.2 has been shown in Section 7.3. Here, only the second part, to recover the controller K, will be shown.

7.5.1 Recovering the Controller

From (7.9), K_h is recovered as

$$K_h = \mathcal{H}_r(\begin{bmatrix} G_{\beta 12}^{-1} & 0 \\ \hat{G}_{\beta 22} G_{\beta 12}^{-1} & G_{\beta 21} \end{bmatrix}, Q_\beta). \tag{7.13}$$

Hence, combining (7.13) with (7.7) and (6.9), the controller K can then be *recovered*, as shown in Figure 7.4(a), as

$$K = \mathcal{H}_r(\begin{bmatrix} I & 0 \\ Z_2 & I \end{bmatrix}, \mathcal{H}_r(\begin{bmatrix} G_{\beta 12}^{-1} & 0 \\ \hat{G}_{\beta 22} G_{\beta 12}^{-1} & G_{\beta 21} \end{bmatrix}, \mathcal{H}_r(\begin{bmatrix} I & 0 \\ Z_1 & I \end{bmatrix} W^{-1}, Q)))$$

$$= \mathcal{H}_r(\begin{bmatrix} G_{\beta 12}^{-1} & 0 \\ G_{\beta 21} F_0 + e^{-sh}\mathcal{F}_u(G_\beta, \gamma^{-2}G_{\beta 11}^\sim)G_{\beta 12}^{-1} & G_{\beta 21} \end{bmatrix} W^{-1}, Q),$$

where Z_1 and W^{-1} are given in (6.5) and (6.6), respectively, with notation changes $B \to B_1$ and $C \to C_1$. According to the formula in Subsection 3.3.1, $\mathcal{F}_u(G_\beta, \gamma^{-2}G_{\beta 11}^\sim)$ has the following realization:

$$\mathcal{F}_u(G_\beta, \gamma^{-2}G_{\beta 11}^\sim) = \left[\begin{array}{cc|c} A & \gamma^{-2}B_1 B_1^* & B_2 \\ -C_1^* C_1 & -A^* & C_1^* \\ \hline -C_2 & \gamma^{-2}B_1^* & 0 \end{array}\right].$$

126   7 The Standard H^∞ Problem

Define
$$F_2 \doteq - \left[\begin{array}{c|c} H & \begin{bmatrix} B_2 \\ C_1^* \end{bmatrix} \\ \hline [-C_2 \ \gamma^{-2}B_1^*] \ \Sigma^{-1} & 0 \end{array} \right],$$

then
$$Z \doteq F_2 + \mathcal{F}_u(G_\beta, \gamma^{-2}G_{\tilde{\beta}11})e^{-sh} = -\pi_h\{\mathcal{F}_u(G_\beta, \gamma^{-2}G_{\tilde{\beta}11})\} \tag{7.14}$$

is an FIR block. As a result,
$$K = \mathcal{H}_r(\begin{bmatrix} I & 0 \\ Z & I \end{bmatrix} \begin{bmatrix} G_{\beta12}^{-1} & 0 \\ G_{\beta21}F_0 - F_2G_{\beta12}^{-1} & G_{\beta21} \end{bmatrix} W^{-1}, Q).$$

Define
$$V^{-1} \doteq \begin{bmatrix} G_{\beta12}^{-1} & 0 \\ G_{\beta21}F_0 - F_2G_{\beta12}^{-1} & G_{\beta21} \end{bmatrix} W^{-1},$$

then
$$K = \mathcal{H}_r(\begin{bmatrix} I & 0 \\ Z & I \end{bmatrix} V^{-1}, Q),$$

as given in (7.10). The simplified structure of K is shown in Figure 7.4(b).

7.5.2 Realization of V^{-1}

$G_{\beta21}F_0 - F_2G_{\beta12}^{-1}$ can be found as

$G_{\beta21}F_0 - F_2G_{\beta12}^{-1}$

$$= \left[\begin{array}{ccc|c} A & \gamma^{-2}B_1B_1^*\Sigma_{21}^* & -\gamma^{-2}B_1B_1^*\Sigma_{11}^* & 0 \\ & A & \gamma^{-2}B_1B_1^* & 0 \\ & -C_1^*C_1 & -A^* & C_1^* \\ \hline -C_2 & \gamma^{-2}B_1^*\Sigma_{21}^* & -\gamma^{-2}B_1^*\Sigma_{11}^* & 0 \end{array} \right]$$

$$+ \left[\begin{array}{cccc|c} A & \gamma^{-2}B_1B_1^* & B_2C_1 & B_2 \\ -C_1^*C_1 & -A^* & C_1^*C_1 & C_1^* \\ 0 & 0 & A + B_2C_1 & B_2 \\ \hline -C_2\Sigma_{22}^* - \gamma^{-2}B_1^*\Sigma_{21}^* & C_2\Sigma_{12}^* + \gamma^{-2}B_1^*\Sigma_{11}^* & 0 & 0 \end{array} \right]$$

$$= \left[\begin{array}{ccc|c} A & 0 & 0 & -\Sigma_{12}^*C_1^* \\ & A & \gamma^{-2}B_1B_1^* & 0 \\ & -C_1^*C_1 & -A^* & C_1^* \\ \hline -C_2 \ \gamma^{-2}B_1^*\Sigma_{21}^* + C_2\Sigma_{22}^* & -\gamma^{-2}B_1^*\Sigma_{11}^* - C_2\Sigma_{12}^* & 0 \end{array} \right]$$

$$+ \left[\begin{array}{cccc|c} A & \gamma^{-2}B_1B_1^* & 0 & 0 \\ -C_1^*C_1 & -A^* & 0 & C_1^* \\ 0 & 0 & A + B_2C_1 & B_2 \\ \hline -C_2\Sigma_{22}^* - \gamma^{-2}B_1^*\Sigma_{21}^* & C_2\Sigma_{12}^* + \gamma^{-2}B_1^*\Sigma_{11}^* & -C_2\Sigma_{22}^* - \gamma^{-2}B_1^*\Sigma_{21}^* & 0 \end{array} \right]$$

$$= \left[\begin{array}{cc|c} A + B_2C_1 & 0 & B_2 \\ 0 & A & -\Sigma_{12}^*C_1^* \\ \hline -\gamma^{-2}B_1^*\Sigma_{21}^* - C_2\Sigma_{22}^* & -C_2 & 0 \end{array} \right].$$

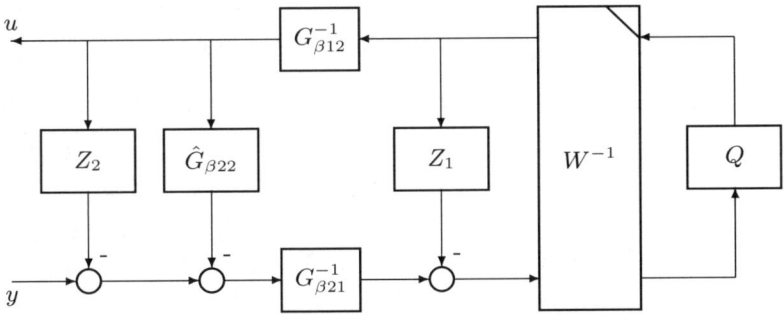

(a) recovered structure

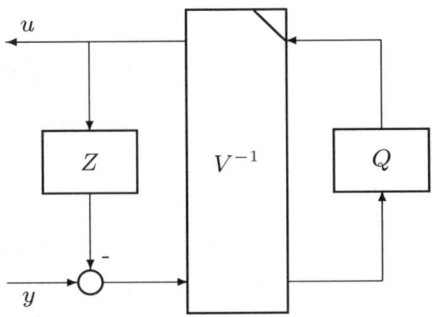

(b) simplified structure

Figure 7.4. Recovering the controller K

Hence,

$$\begin{bmatrix} G_{\beta12}^{-1} & 0 \\ G_{\beta21}F_0 - F_2 G_{\beta12}^{-1} & G_{\beta21} \end{bmatrix} = \left[\begin{array}{cc|cc} A+B_2C_1 & 0 & B_2 & 0 \\ 0 & A & -\Sigma_{12}^*C_1^* & B_1 \\ \hline C_1 & 0 & I & 0 \\ -\gamma^{-2}B_1^*\Sigma_{21}^* - C_2\Sigma_{22}^* & -C_2 & 0 & I \end{array} \right],$$

and V^{-1} is obtained as

$$V^{-1} = \begin{bmatrix} G_{\beta12}^{-1} & 0 \\ G_{\beta21}F_0 - F_2 G_{\beta12}^{-1} & G_{\beta21} \end{bmatrix} W^{-1}$$

$$= \left[\begin{array}{cccc|cc} A+B_2C_1 & 0 & -B_2C_1 & & B_2 & 0 \\ 0 & A & \Sigma_{12}^*C_1^*C_1 + \gamma^{-2}B_1B_1^*\Sigma_{21}^* & & -\Sigma_{12}^*C_1^* & B_1 \\ 0 & 0 & A & & \Sigma_{12}\Sigma_{22}^{-1}C_1^* & -\Sigma_{22}^{-*}B_1 \\ \hline C_1 & 0 & -C_1 & & I & 0 \\ -\gamma^{-2}B_1^*\Sigma_{21}^* - C_2\Sigma_{22}^* & -C_2 & \gamma^{-2}B_1^*\Sigma_{21}^* & & 0 & I \end{array} \right]$$

$$= \begin{bmatrix} A+B_2C_1 & 0 & 0 & B_2 - \Sigma_{12}\Sigma_{22}^{-1}C_1^* & \Sigma_{22}^{-*}B_1 \\ 0 & A & 0 & 0 & 0 \\ 0 & 0 & A & \Sigma_{12}\Sigma_{22}^{-1}C_1^* & -\Sigma_{22}^{-*}B_1 \\ \hline C_1 & 0 & 0 & I & 0 \\ -\gamma^{-2}B_1^*\Sigma_{21}^* - C_2\Sigma_{22}^* & -C_2 & 0 & 0 & I \end{bmatrix}$$

$$= \begin{bmatrix} A+B_2C_1 & B_2 - \Sigma_{12}\Sigma_{22}^{-1}C_1^* & \Sigma_{22}^{-*}B_1 \\ \hline C_1 & I & 0 \\ -\gamma^{-2}B_1^*\Sigma_{21}^* - C_2\Sigma_{22}^* & 0 & I \end{bmatrix},$$

where the following equality given in Lemma 3.14 is used:

$$\Sigma_{22}^* A - A\Sigma_{22}^* + \gamma^{-2}B_1 B_1^* \Sigma_{21}^* + \Sigma_{12}^* C_1^* C_1 = 0.$$

7.6 Summary and Notes

This chapter derives a complete solution to the standard H^∞ problem of rational plants with a single I/O delay. It is solved by splitting the original problem to a delay-free problem and a one-block problem (OP$_h$) and then OP$_h$ is further reduced to an extended Nehari problem (ENP$_h$), mainly using a new interpretation of Kimura's theorem [58, Lemma 7.1]. Using the solution to the ENP$_h$ proposed in the previous chapter, the H^∞ controllers for the the one-block problem and the standard H^∞ control problem are recovered.

The idea used in this chapter may be useful to solve H^∞ control problem for systems with multiple delays. The suboptimal controller incorporates an FIR block. This FIR block can be obtained by combining two FIR blocks using two steps. This may be a useful hint to solve the multi-delay problem. It has been shown that the standard H^∞ control problem for plants having constraints can be solved in a clear and simple way by reducing the original problem to a nonconstrained problem and a one-block problem with constraints. Once the one-block problem is solved, the original problem is solved. The fact that the inverse of the lower-right block of $\mathcal{C}_r(P)\begin{bmatrix} I & 0 \\ Z & I \end{bmatrix}V^{-1}$ is also stable may well be useful for the robust stability analysis of time-delay systems, which is a very active field [43, 44, 164].

8

A Transformed Standard H^∞ Problem

It has been shown in the previous chapter that the standard H^∞ problem of time-delay systems is very complicated. In this chapter,[1] a transformation is presented to solve the standard H^∞ problems of dead-time systems similar to the finite-dimensional situations. With some trade-off of the performance, the following advantages are obtained: (i) The controller has a quite simple and transparent structure; (ii) There are no any *additional* hidden modes in the Smith predictor. As a result, the practical significance of the approach is obvious.

8.1 Introduction

Dead-time systems are systems in which the action of control inputs takes a certain time before it affects the measured outputs. The typical dead-time systems are those with input delays. It is well known that it is very difficult to control such systems. The Smith predictor [126] was the first effective way to control such systems and many modified Smith predictors have been proposed [113]. In recent years, many researchers have been interested in the optimal control of dead-time systems, especially H_∞ control, *i.e.*, to find a controller to internally stabilise the system (if so, called an admissible controller) and to minimise the H_∞-norm of the transfer matrix from the external input signals (such as noises, disturbances and reference signals) to the output signals (such as controlled signals and tracking errors).

With the help of the modified Smith predictor, many robust control problems for dead-time systems, such as robust stability, tracking and model-matching and input sensitivity minimisation, can be solved as in finite-dimensional situations [80, 104]. However, the sensitivity minimisation, the mixed sensitivity minimisation and/or the standard H_∞ control problems cannot be solved in this way. Very recently, notable results were presented in

---

[1] Portions reprinted from [157, 161], with permission from Elsevier.

[80, 85, 100, 134, 160, 162] using different methods. The results in [80, 85, 162], which are formulated in the form of a modified Smith predictor, are quite elegant and the ideas are very tricky, but the controllers are too involved. Specifically, the Smith predictor involved is complex and, even more, relates to the performance level γ. Hence, there exist problems in applying the method to systems with long dead time, as pointed out by Meinsma and Zwart [80]. See also Chapter 10. Another disadvantage is that the predictor (see, F_{stab} in [80, Theorem 5.3], $\Delta_{\alpha,\infty}$ in [85, Theorem 2] or $\Delta(s)$ in [162, Theorem 2]) always includes *additional* unstable hidden modes, even with stable plants, because the hidden modes are the eigenvalues of a Hamiltonian matrix. As can be seen from Chapter 11, the implementation of such a predictor is not trivial.

In this chapter, a transformation of the closed-loop transfer matrix of dead-time systems is proposed to simplify the problem. In fact, it is a new H_∞ performance evaluation scheme for dead-time systems. With this transformation, all robust control problems can be solved as in the finite-dimensional situations. The controller obtained has a simple and transparent structure incorporating a modified Smith predictor. The resulting Smith predictor depends only on the real plant and is independent of the performance level γ and of the performance evaluation scheme. There do not exist any *additional* hidden modes in the predictor and, hence, it is easy to implement. The cost to pay for these advantages is some performance loss from the conventional H_∞ control point of view. When the dead-time h becomes 0, the transformation becomes null. Hence, it can be regarded as an extension of the conventional performance evaluation scheme for dead-time systems. This transformation does not affect the original system and is easy to apply because it, *virtually*, subtracts an finite-impulse-response (FIR) block from the original system.

The rest of this chapter is organised as follows. The transformation (or the new performance evaluation scheme) is presented in Section 8.2; the solution to H_∞ control of dead-time systems with an input/output delay is given in Section 8.3 and an example is given in Section 8.4.

8.2 The Transformation

The general control setup for dead-time systems with a input or output delay is shown in Figure 8.1, where

$$P(s) = \begin{bmatrix} P_{11}(s) & P_{12}(s) \\ P_{21}(s) & P_{22}(s) \end{bmatrix}.$$

The closed-loop transfer matrix from w to z is

$$T_{zw}(s) = P_{11} + e^{-sh}P_{12}K(I - e^{-sh}P_{22}K)^{-1}P_{21}.$$

This means that there exists an instantaneous response through the path P_{11} without any delay. A clearer equivalent structure is shown in Figure 8.2. It is not difficult to recognise that, during the period $t = 0 \sim h$ after w is applied, the output z is *not* controllable (*i.e.*, not changeable by the control action) and is *only* determined by P_{11} (and, of course, w). However, the response during this period may dominate the system performance index. This means the performance index is likely dominated by a response that cannot be controlled. This is not what is expected. It does not make sense to include such an uncontrollable item in the performance index. Hence, this part should be excluded from the response when evaluating the system performance. In general, it is impossible to eliminate the instantaneous response by simply introducing a suitable rational P_{11}.

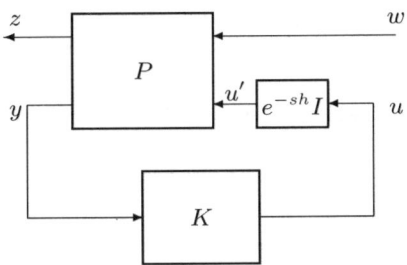

Figure 8.1. General control setup for dead-time systems

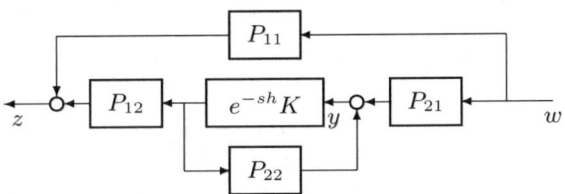

Figure 8.2. An equivalent structure

Define an FIR block

$$Z_1(s) = \tau_h\{P_{11}\} = P_{11}(s) - \tilde{P}_{11}(s)e^{-sh},$$

which is exactly the uncontrollable part in $T_{zw}(s)$. Subtract it from the feedforward path P_{11}, as shown in Figure 8.3, then

$$T_{zw}(s) = Z_1(s) + T_{z'w}(s), \tag{8.1}$$

where

$$T_{z'w}(s) = e^{-sh}\{\tilde{P}_{11} + P_{12}K(I - e^{-sh}P_{22}K)^{-1}P_{21}\}.$$

The fact shown in (8.1) was recognised in [90] to parameterise all stabilising dead-time controllers. There is no instantaneous response in $T_{z'w}(s)$. The only difference between $T_{z'w}(s)$ and $T_{zw}(s)$ is the FIR block $Z_1(s)$, which is not a part of the real control system but an artificial part introduced into the performance evaluation scheme. It is this FIR block that makes the control problems so complex and difficult to solve. As shown in [85], the achievable minimal performance index $\|T_{zw}\|_\infty$ is larger than $\|Z_1\|_\infty$. Hence,

$$\|Z_1\|_\infty < \|T_{zw}\|_\infty \leq \|Z_1\|_\infty + \|T_{z'w}\|_\infty. \tag{8.2}$$

In this chapter, $\|T_{z'w}\|_\infty$, instead of $\|T_{zw}\|_\infty$, is optimised. Once $\|T_{z'w}\|_\infty$ is minimised, the achievable performance $\|T_{zw}\|_\infty$ is not larger than $\|Z_1\|_\infty + \|T_{z'w}\|_\infty$ and so is the optimal performance. Using the inequality (8.2), one can estimate how far the suboptimal controller obtained is away from the optimal controller.

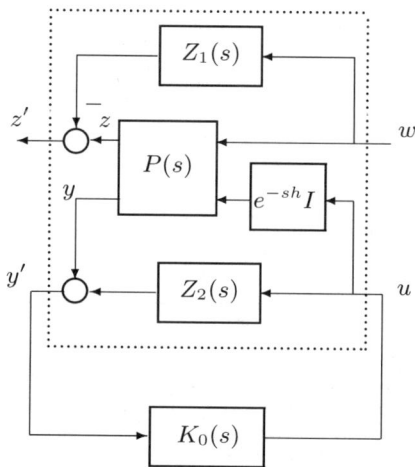

Figure 8.3. Graphic interpretation of the transformation

8.3 Solution

Assume that the realization of the rational part of the generalised process in Figure 8.1 is taken to be of the form

$$P(s) = \left[\begin{array}{c|cc} A & B_1 & B_2 \\ \hline C_1 & D_{11} & D_{12} \\ C_2 & D_{21} & D_{22} \end{array}\right]$$

and the following standard assumptions hold:

(A1) (A, B_2) is stabilisable and (C_2, A) is detectable;

(A2) $\begin{bmatrix} A - j\omega I & B_2 \\ C_1 & D_{12} \end{bmatrix}$ has full column rank $\forall \omega \in R$;

(A3) $\begin{bmatrix} A - j\omega I & B_1 \\ C_2 & D_{21} \end{bmatrix}$ has full row rank $\forall \omega \in R$;

(A4) $D_{12}^* D_{12} = I$ and $D_{21} D_{21}^* = I$.

Assumption (A4) is made to simplify the exposition. In fact, only the nonsingularity of the matrices $D_{12}^* D_{12}$ and $D_{21} D_{21}^*$ is required.

Consider a Smith predictor-type controller

$$K(s) = K_0(s)(I - Z_2(s)K_0(s))^{-1},$$

as shown in Figure 8.3, in which the predictor is designed to be

$$Z_2(s) = \pi_h\{P_{22}\} = \left[\begin{array}{c|c} A & B_2 \\ \hline C_2 e^{-Ah} & 0 \end{array}\right] - e^{-sh} \left[\begin{array}{c|c} A & B_2 \\ \hline C_2 & D_{22} \end{array}\right],$$

then the system can be re-formulated as

$$\begin{bmatrix} z'' \\ y' \end{bmatrix} = \widetilde{P}(s) \begin{bmatrix} w \\ u \end{bmatrix}$$
$$u = K_0(s) y'$$

with $e^{-sh} z'' \doteq z'$, where

$$\widetilde{P}(s) = \left[\begin{array}{c|cc} A & e^{Ah} B_1 & B_2 \\ \hline C_1 & 0 & D_{12} \\ C_2 e^{-Ah} & D_{21} & 0 \end{array}\right]$$

is free of dead time (but delay-dependent) and K_0 is called the main controller. The closed-loop transfer function from w to z' is

$$T_{z'w}(s) = e^{-sh} T_{z''w}(s) = e^{-sh} \mathcal{F}_l(\widetilde{P}(s), K_0(s)). \tag{8.3}$$

Hence, the H_∞ control problem

$$\|T_{z'w}(s)\|_\infty < \gamma$$

is converted to

$$\left\|\mathcal{F}_l(\widetilde{P}(s), K_0(s))\right\|_\infty < \gamma.$$

This is a finite-dimensional problem, which can be solved using known results. Since this is for the general setup of systems with an input/output delay, all the H_∞ control problems (such as robust stability, tracking and model-matching, input sensitivity minimisation, output sensitivity minimisation, mixed sensitivity minimisation and the standard H_∞ control problem *etc.*) can be solved similarly to the finite-dimensional situations.

The solution, which is given in Theorem 8.1 below, involves two Hamiltonian matrices:

$$H_h = \begin{bmatrix} A & \gamma^{-2} e^{Ah} B_1 B_1^* e^{A^* h} \\ -C_1^* C_1 & -A^* \end{bmatrix} - \begin{bmatrix} B_2 \\ -C_1^* D_{12} \end{bmatrix} \begin{bmatrix} D_{12}^* C_1 & B_2^* \end{bmatrix},$$

$$J_h = \begin{bmatrix} A^* & \gamma^{-2} C_1^* C_1 \\ -e^{Ah} B_1 B_1^* e^{A^* h} & -A \end{bmatrix} - \begin{bmatrix} e^{-A^* h} C_2^* \\ -e^{Ah} B_1 D_{21}^* \end{bmatrix} \begin{bmatrix} D_{21} B_1^* e^{A^* h} & C_2 e^{-Ah} \end{bmatrix}.$$

Theorem 8.1. *There exists a main controller such that* $\|T_{z'w}(s)\|_\infty < \gamma$ *in Figure 8.3 iff the following three conditions hold:*
(i) $H_h \in dom(Ric)$ *and* $X = Ric(H_h) \geq 0$;
(ii) $J_h \in dom(Ric)$ *and* $Y = Ric(J_h) \geq 0$;
(iii) $\rho(XY) < \gamma^2$.
Moreover, when the conditions hold, one such main controller is

$$K_0(s) = \left[\begin{array}{c|c} A_h & -L_h \\ \hline F_h\Psi_h & 0 \end{array}\right],$$

where

$$A_h = A + L_h C_2 e^{-Ah} + \gamma^{-2} Y C_1^* C_1 + \left(B_2 + \gamma^{-2} Y C_1^* D_{12}\right) F_h \Psi_h,$$

$$F_h = -(B_2^* X + D_{12}^* C_1),$$

$$L_h = -(Y e^{-A^* h} C_2^* + e^{Ah} B_1 D_{21}^*),$$

$$\Psi_h = (I - \gamma^{-2} Y X)^{-1}.$$

Furthermore, the set of all admissible main controllers such that $\|T_{z'w}(s)\|_\infty < \gamma$ *can be parameterised as*

$$K_0(s) = \mathcal{F}_l(M(s), Q(s)),$$

where

$$M(s) = \left[\begin{array}{c|cc} A_h & -L_h & B_2 + \gamma^{-2} Y C_1^* D_{12} \\ \hline F_h \Psi_h & 0 & I \\ -\left(C_2 e^{-Ah} + \gamma^{-2} D_{21} B_1^* e^{A^* h} X\right)\Psi_h & I & 0 \end{array}\right]$$

and $Q(s) \in H_\infty$, $\|Q(s)\|_\infty < \gamma$.

Proof. First of all, check if $\tilde{P}(s)$ meets the standard assumptions (A1–A4).
 (A1) $(C_2 e^{-Ah}, A)$ is detectable because $A + e^{Ah} L C_2 e^{-Ah} \sim A + LC_2$ and (C_2, A) is detectable;
 (A3) $\left[\begin{smallmatrix} A - j\omega I & e^{Ah} B_1 \\ C_2 e^{-Ah} & D_{21} \end{smallmatrix}\right]$ has full row rank $\forall \omega \in R$ because $\left[\begin{smallmatrix} A - j\omega I & e^{Ah} B_1 \\ C_2 e^{-Ah} & D_{21} \end{smallmatrix}\right] \sim \left[\begin{smallmatrix} A - j\omega I & B_1 \\ C_2 & D_{21} \end{smallmatrix}\right]$ has full row rank $\forall \omega \in R$.
 The assumptions (A2) and (A4) remain unchanged. Hence, $\tilde{P}(s)$ meets all the standard assumptions. Theorem 5.1 in [39] can be directly used to solve the problem. Substitute $\tilde{P}(s)$ into the theorem, then the above result can be obtained with ease. □

Remark 8.1. Under this transformation, neither $D_{11} \neq 0$ nor $D_{22} \neq 0$ makes the problem more complicated. In fact, when $h = 0$, $Z_2(s) = -D_{22}$ is the common controller transformation to make $D_{22} = 0$ [180].

8.4 A Numerical Example

Consider the example studied in [80], which is shown in Figure 8.4 with

$$P_r(s) = \frac{1}{s-1} \text{ and } P_{rd}(s) = \frac{s-1}{s+1},$$

$$W_1(s) = \frac{2(s+1)}{10s+1} \text{ and } W_2(s) = \frac{0.2(s+1.1)}{s+1}.$$

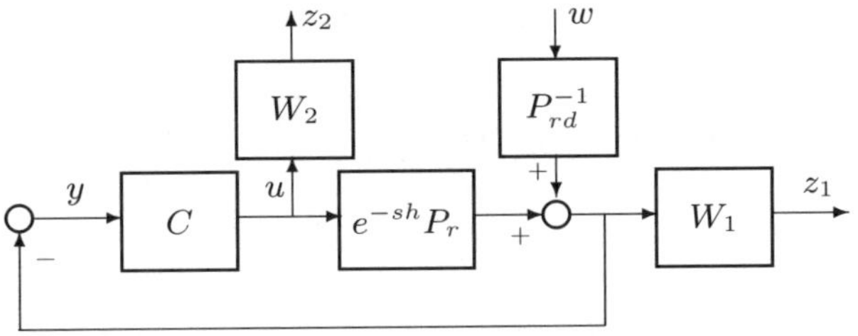

Figure 8.4. Setup for mixed sensitivity minimisation

This problem can be arranged as a standard problem with

$$P(s) = \begin{bmatrix} \frac{2(s+1)^2}{(10s+1)(s-1)} & \frac{2(s+1)}{(10s+1)(s-1)} \\ 0 & \frac{0.2(s+1.1)}{s+1} \\ \hline -\frac{s+1}{s-1} & -\frac{1}{s-1} \end{bmatrix}.$$

Here, $P_{22} = -\frac{1}{s-1}$, $P_{11} = \begin{bmatrix} \frac{2(s+1)^2}{(10s+1)(s-1)} & 0 \end{bmatrix}^T$. $W_2(s)$ is delayed to be $W_2 e^{-sh}$, which does not affect the result.

The predictor is designed to be

$$Z_2(s) = -\frac{e^{-h} - e^{-sh}}{s-1}.$$

There is no additional hidden pole. However, the predictor

$$F_{stab}(s) = -\frac{10.75s^2 + 0.245s - 0.3298}{(s-1)(12.03s^2+1)} + \frac{13.16s^2 - 0.1316}{(s-1)(12.03s^2+1)} e^{-sh}$$

obtained using the approach in [80] is much more complicated and has two additional hidden poles $s_{1,2} = \pm 0.2883j$.

The achievable performance is listed in Table 8.1 for different delays h. The performance does not degrade much to pay for the advantages obtained. The frequency responses of both cases are shown in Figure 8.5. Although the H_∞-norm achieved is slightly larger than that achieved by the method of Meinsma

136   8 A Transformed Standard H^∞ Problem

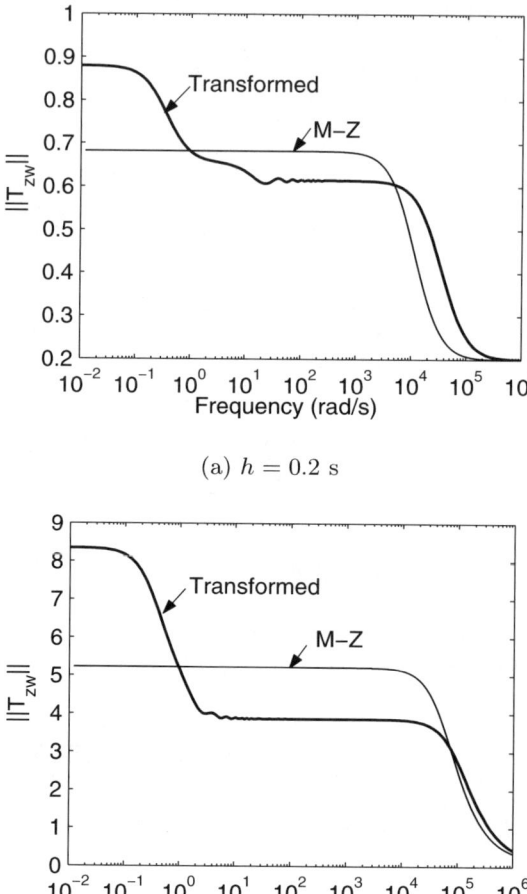

(a) $h = 0.2$ s

(b) $h = 2$ s

Figure 8.5. Comparison of $\|T_{zw}(j\omega)\|$

Table 8.1. Performance comparison

h	M-Z's $\|T_{zw}(s)\|_\infty$	Transformed $\|T_{z'w}(s)\|_\infty$	Transformed $\|T_{zw}(s)\|_\infty$
0.2	0.68	0.58	0.88
2	5.22	3.86	8.34
5	109.5*	78.63	183.7

*Note: For long dead time h, e.g., $h > 3.2$ s, some matrices are close to singular or badly scaled; the result of M-Z is likely inaccurate.

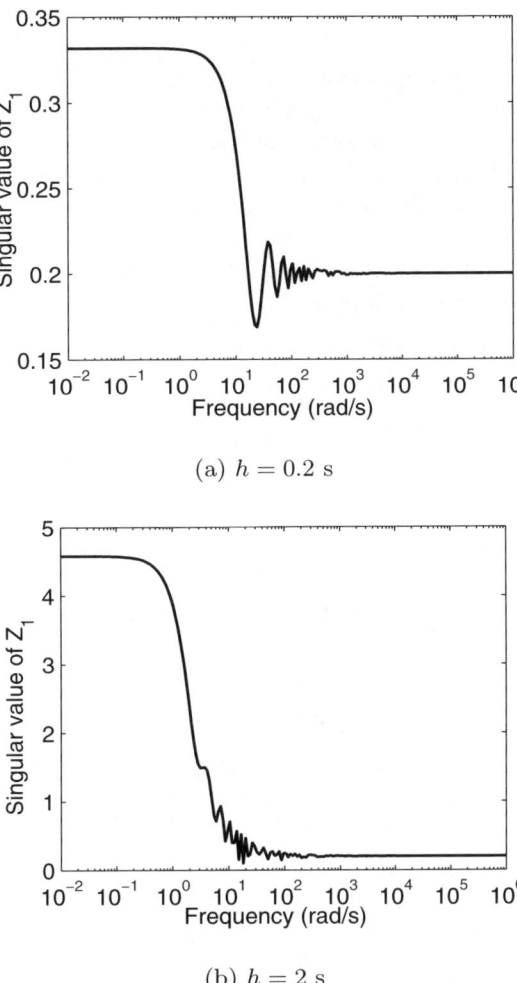

Figure 8.6. Singular value plots of $Z_1(s)$

and Zwart (denoted M-Z in the figures), the transformed method gives better performance over quite a broad frequency band ranging from 1 rad/s to about 10,000 rad/s. From the engineering point of view, this solution is much better. The singular value plot of the uncontrollable part $Z_1(s)$ is shown in Figure 8.6. It is clear that the high gain of T_{zw} at low frequencies is primarily due to the large singular value of $Z_1(s)$ at low frequencies.

8.5 Summary

In this chapter, a transformation (in fact, a new performance evaluation scheme) is introduced for the H_∞ control of dead-time systems. This considerably simplifies the H_∞ control problem of dead-time systems and has many advantages. In particular, the practical significance is obvious. Since this is not a solution to the original H_∞ control problem of time-delay systems there may exist some performance loss from the conventional H_∞ control or mathematical point of view. An inequality can be used to estimate how far the suboptimal controller is away from the optimal controller.

9
2DOF Controller Parameterisation

In this chapter, the co-prime factorisation of all stabilising controllers is presented and then the controller is realized as a two-degree-of-freedom structure. One degree-of-freedom $F(s)$ is chosen to meet the desired set-point response and the other, the free parameter $Q(s)$, is chosen as a compromise between disturbance response and robustness. Furthermore, the controller is re-configured in chain-scattering representation. With this structure, which is symmetrical for the process and the disturbance degree-of-freedom $Q(s)$, the two degrees-of-freedom and the differences between controllers for processes with and without dead time can clearly be seen. It is also shown that the subideal disturbance response can be obtained with a suitable choice of $Q(s)$. As a special case, the method is applied to integral processes with dead time.

9.1 Parameterisation of the Controller

Consider the following processes with a single delay $h \geq 0$:
$$G_p(s) = P(s)e^{-hs}, \tag{9.1}$$
where $P(s)$ is a rational transfer matrix with a minimal realization
$$P(s) = \left[\begin{array}{c|c} A & B \\ \hline C & 0 \end{array}\right] = C(sI - A)^{-1}B.$$
Introduce a delay-free transfer matrix
$$P_0(s) = \left[\begin{array}{c|c} A & B \\ \hline Ce^{-Ah} & 0 \end{array}\right] = Ce^{-Ah}(sI - A)^{-1}B \tag{9.2}$$
and assume that P_0 can be co-primely factorised as
$$P_0 = N_0 M^{-1} = \tilde{M}^{-1}\tilde{N}_0 \tag{9.3}$$

9 2DOF Controller Parameterisation

with[1] $M, N_0, \tilde{M}, \tilde{N}_0 \in H_\infty$. Then there exist rational transfer matrices $X_0, Y, \tilde{X}_0, \tilde{Y} \in H_\infty$ such that the following *Bezout identity* holds:

$$\begin{bmatrix} \tilde{X}_0 & \tilde{Y} \\ -\tilde{N}_0 & \tilde{M} \end{bmatrix} \begin{bmatrix} M & -Y \\ N_0 & X_0 \end{bmatrix} = \begin{bmatrix} I & 0 \\ 0 & I \end{bmatrix}. \tag{9.4}$$

The set of all proper controllers internally stabilising the delay-free processes $P_0(s)$ is given by

$$K_0 = (-Y + MQ)(X_0 + N_0Q)^{-1} = \mathcal{H}_r\left(\begin{bmatrix} M & -Y \\ N_0 & X_0 \end{bmatrix}, Q\right), \tag{9.5}$$

or

$$\tilde{K}_0 = (\tilde{X}_0 + Q\tilde{N}_0)^{-1}(-\tilde{Y} + Q\tilde{M}) = \mathcal{H}_l\left(\begin{bmatrix} \tilde{X}_0 & \tilde{Y} \\ -\tilde{N}_0 & \tilde{M} \end{bmatrix}, Q\right) \tag{9.6}$$

for any $Q \in H_\infty$.

Since $P(s)$ and $P_0(s)$ have the same denominator $\det(sI - A)$, $P(s)$ can be co-primely factorised over H_∞ as

$$P(s) = NM^{-1} = \tilde{M}^{-1}\tilde{N} \tag{9.7}$$

and the process can be factorised as

$$G_p(s) = (Ne^{-hs})M^{-1} = \tilde{M}^{-1}(\tilde{N}e^{-hs}).$$

Define a predictor as[2]

$$\begin{aligned} Z(s) &= P_0(s) - P(s)e^{-hs} \\ &= N_0M^{-1} - (Ne^{-hs})M^{-1} \\ &= \tilde{M}^{-1}\tilde{N}_0 - \tilde{M}^{-1}(\tilde{N}e^{-hs}). \end{aligned}$$

Then, $Z(s)$ is stable because

$$\begin{aligned} Z(s) &= Ce^{-Ah}(sI - A)^{-1}B - C(sI - A)^{-1}Be^{-hs} \\ &= Ce^{-Ah}\left(I - e^{-h(sI-A)}\right)(sI - A)^{-1}B \end{aligned}$$

is an entire function of s (actually, it has a finite impulse response with support on $[0, h]$) and belongs to H_∞ [90, 113].

Theorem 9.1. *Let $M, N_0, \tilde{M}, \tilde{N}_0, X_0, Y, \tilde{X}_0, \tilde{Y}$ be a double co-prime factorisation (9.3, 9.4) of $P_0(s)$ over H_∞, then the set of all proper controllers internally stabilising the processes with dead time (9.1) is parameterised as*

$$K = \mathcal{H}_r\left(\begin{bmatrix} I & 0 \\ -Z & I \end{bmatrix}, K_0\right) \tag{9.8}$$

---
[1] The argument s of a transfer matrix is often omitted for simplicity.
[2] The choice of Z is not unique. See Chapter 10.

9.1 Parameterisation of the Controller

or as

$$K = \mathcal{H}_l \left(\begin{bmatrix} I & 0 \\ Z & I \end{bmatrix}, \tilde{K}_0 \right), \tag{9.9}$$

where K_0 and $\tilde{K}_0$, as given in (9.5) and (9.6) respectively, are the set of all stabilising controllers for $P_0(s)$.

Proof. Since

$$\begin{bmatrix} I & 0 \\ -Z & I \end{bmatrix} \begin{bmatrix} M & -Y \\ N_0 & X_0 \end{bmatrix} = \begin{bmatrix} M & -Y \\ Ne^{-hs} & X \end{bmatrix}$$

and

$$\begin{bmatrix} \tilde{X}_0 & \tilde{Y} \\ -\tilde{N}_0 & \tilde{M} \end{bmatrix} \begin{bmatrix} I & 0 \\ Z & I \end{bmatrix} = \begin{bmatrix} \tilde{X} & \tilde{Y} \\ -\tilde{N}e^{-hs} & \tilde{M} \end{bmatrix},$$

where $X = X_0 + ZY$ and $\tilde{X} = \tilde{X}_0 + \tilde{Y}Z$, then

$$\begin{bmatrix} \tilde{X} & \tilde{Y} \\ -\tilde{N}e^{-hs} & \tilde{M} \end{bmatrix} \begin{bmatrix} M & -Y \\ Ne^{-hs} & X \end{bmatrix}$$
$$= \begin{bmatrix} \tilde{X}_0 & \tilde{Y} \\ -\tilde{N}_0 & \tilde{M} \end{bmatrix} \begin{bmatrix} I & 0 \\ Z & I \end{bmatrix} \begin{bmatrix} I & 0 \\ -Z & I \end{bmatrix} \begin{bmatrix} M & -Y \\ N_0 & X_0 \end{bmatrix}$$
$$= \begin{bmatrix} \tilde{X}_0 & \tilde{Y} \\ -\tilde{N}_0 & \tilde{M} \end{bmatrix} \begin{bmatrix} M & -Y \\ N_0 & X_0 \end{bmatrix}$$
$$= \begin{bmatrix} I & 0 \\ 0 & I \end{bmatrix}.$$

This means that the above equation is a double co-prime factorisation of the processes with dead time. According to the Theorem 11.6 in [180, Chapter 11], the set of all proper controllers achieving internal stability of the processes with dead time is parameterised either by

$$K(s) = (-Y + MQ)(X + Ne^{-hs}Q)^{-1}$$
$$= \mathcal{H}_r \left(\begin{bmatrix} M & -Y \\ Ne^{-hs} & X \end{bmatrix}, Q \right)$$
$$= \mathcal{H}_r \left(\begin{bmatrix} I & 0 \\ -Z & I \end{bmatrix} \begin{bmatrix} M & -Y \\ N_0 & X_0 \end{bmatrix}, Q \right)$$
$$= \mathcal{H}_r \left(\begin{bmatrix} I & 0 \\ -Z & I \end{bmatrix}, K_0 \right)$$

or by

$$K(s) = (\tilde{X} + Q\tilde{N}e^{-hs})^{-1}(-\tilde{Y} + Q\tilde{M})$$
$$= \mathcal{H}_l \left(\begin{bmatrix} \tilde{X} & \tilde{Y} \\ -\tilde{N}e^{-hs} & \tilde{M} \end{bmatrix}, Q \right)$$
$$= \mathcal{H}_l \left(\begin{bmatrix} \tilde{X}_0 & \tilde{Y} \\ -\tilde{N}_0 & \tilde{M} \end{bmatrix} \begin{bmatrix} I & 0 \\ Z & I \end{bmatrix}, Q \right)$$
$$= \mathcal{H}_l \left(\begin{bmatrix} I & 0 \\ Z & I \end{bmatrix}, \tilde{K}_0 \right)$$

with K_0 and $\tilde{K}_0$ given in (9.5) and (9.6) for any $Q \in H_\infty$ since $\lim_{s \to \infty}(I + X^{-1}Ne^{-hs}Q) = I \neq 0$ and $\lim_{s \to \infty}(I + Q\tilde{N}e^{-hs}\tilde{X}^{-1}) = I \neq 0$. □

Remark 9.1. $Z = P_0(s) - P(s)e^{-hs}$ is a modified Smith predictor. When $P(s)$ is stable, $P_0(s)$ may be chosen as $P(s)$ and then Z is the well-known Smith predictor. Thus, $N_0 = N$, $\tilde{N}_0 = \tilde{N}$.

Remark 9.2. If $P(s)$ is unstable, then the implementation of $Z(s)$ is not trivial. See Part II for more details.

Remark 9.3. A similar idea was used in [19].

9.2 Two-degree-of-freedom Realization of the Controller

9.2.1 Control Structure

There are many realizations of the controllers given in (9.8) and (9.9). A scheme with two-degrees-of-freedom for (9.9) is shown in Figure 9.1(a), where $F(s)$ is a pre-filter. It is easy to see that the two degrees-of-freedom are $F(s)$ and $Q(s)$. If $h = 0$, then $P_0(s) = P(s)$ and $Z = 0$. The block $\begin{bmatrix} I & 0 \\ Z & I \end{bmatrix}$ becomes an identity matrix and can be neglected. In other words, the only effect of dead time h on the control system is to insert a block $\begin{bmatrix} I & 0 \\ Z & I \end{bmatrix}$ between the delay-free controller block $\begin{bmatrix} \tilde{X}_0 & \tilde{Y} \\ -\tilde{N}_0 & \tilde{M} \end{bmatrix}$ and the process $P(s)e^{-hs}$.

According to this structure,

$$a = F \cdot R + Q \cdot b \qquad (9.10)$$

and

$$\begin{bmatrix} a \\ b \end{bmatrix} = \begin{bmatrix} \tilde{X}_0 & \tilde{Y} \\ -\tilde{N}_0 & \tilde{M} \end{bmatrix} \begin{bmatrix} U_1 \\ Y_1 \end{bmatrix}. \qquad (9.11)$$

Substituting (9.11) into (9.10),

$$F \cdot R + Q(-\tilde{N}_0 U_1 + \tilde{M}Y_1) = \tilde{X}_0 U_1 + \tilde{Y}Y_1.$$

Hence,

$$U_1 = (\tilde{X}_0 + Q\tilde{N}_0)^{-1}\left((-\tilde{Y} + Q\tilde{M})Y_1 + F \cdot R\right). \qquad (9.12)$$

This gives the equivalent structure in the classical representation shown in Figure 9.1(b).

9.2 Two-degree-of-freedom Realization of the Controller

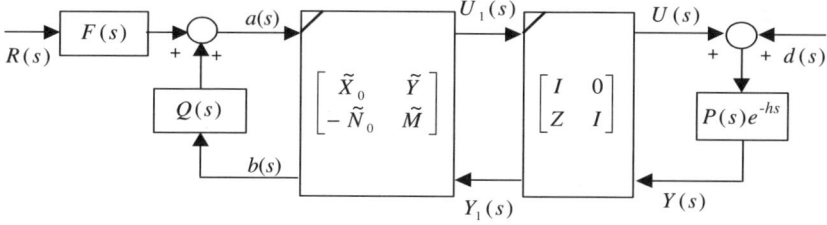

(a) in the chain-scattering representation

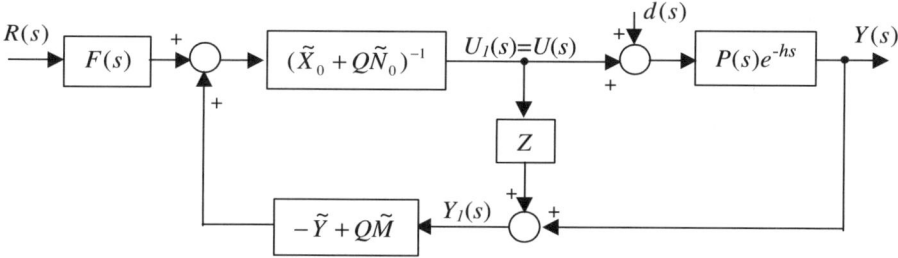

(b) the equivalent classical representation

Figure 9.1. Parameterised control structure with 2DOF

9.2.2 Set-point Response

Assume $d(s) = 0$. Then in the nominal case,

$$Y_1 = P_0 U_1$$

and the transfer function from $R(s)$ to $U_1(s) = U(s)$ is

$$G_{ur} = (I - \tilde{K}_0 P_0)^{-1} (\tilde{X}_0 + Q\tilde{N}_0)^{-1} F.$$

Hence, the set-point response, *i.e.*, the transfer function from the reference signal $R(s)$ to the output $Y(s)$, is

$$\begin{aligned} G_{yr}(s) &= P(s)e^{-hs} G_{ur}(s) \\ &= P(s)e^{-hs} \left(I - \tilde{K}_0(s)\tilde{M}^{-1}\tilde{N}_0\right)^{-1} \cdot (\tilde{X}_0 + Q\tilde{N}_0)^{-1} F(s) \\ &= P(s)(\tilde{X}_0 + \tilde{Y}\tilde{M}^{-1}\tilde{N}_0)^{-1} F(s) e^{-hs} \\ &= NM^{-1}(\tilde{X}_0 + \tilde{Y}N_0 M^{-1})^{-1} F(s) e^{-hs} \\ &= N(s) F(s) e^{-hs}. \end{aligned}$$

This is stable as long as $F(s)$ is stable. Moreover, it has nothing to do with the free parameter $Q(s)$.

9.2.3 Disturbance Response

Assume $R(s) = 0$. Then according to (9.12),

$$U = U_1 = (\tilde{X}_0 + Q\tilde{N}_0)^{-1}(-\tilde{Y} + Q\tilde{M})Y_1 = \tilde{K}_0 Y_1.$$

On the other hand, according to Figure 9.1(b),

$$Y_1 = P_0 U + Pe^{-hs} \cdot d.$$

Hence,

$$U = \tilde{K}_0 (I - P_0 \tilde{K}_0)^{-1} Pe^{-hs} \cdot d$$

and the output is

$$\begin{aligned}
Y &= Pe^{-hs}(U + d) \\
&= Pe^{-hs}(\tilde{K}_0(I - P_0\tilde{K}_0)^{-1} Pe^{-hs} + I)d \\
&= Pe^{-hs}((I - \tilde{K}_0 P_0)^{-1}\tilde{K}_0 Pe^{-hs} + I)d \\
&= Pe^{-hs}(I - \tilde{K}_0 P_0)^{-1}(I - \tilde{K}_0 Z)d.
\end{aligned}$$

In other words, the disturbance response is

$$\begin{aligned}
G_{yd}(s) &= Pe^{-hs}(I - \tilde{K}_0 P_0)^{-1}(I - \tilde{K}_0 Z) \\
&= Pe^{-hs}\mathcal{H}_l\left(\begin{bmatrix} I & -I \\ P_0 & -Z \end{bmatrix}, \tilde{K}_0\right) \\
&= Pe^{-hs}\mathcal{H}_l\left(\begin{bmatrix} I & -I \\ P_0 & -Z \end{bmatrix}, \mathcal{H}_l\left(\begin{bmatrix} \tilde{X}_0 & \tilde{Y} \\ -\tilde{N}_0 & \tilde{M} \end{bmatrix}, Q\right)\right) \\
&= Pe^{-hs}\mathcal{H}_l\left(\begin{bmatrix} \tilde{X}_0 & \tilde{Y} \\ -\tilde{N}_0 & \tilde{M} \end{bmatrix}\begin{bmatrix} I & -I \\ P_0 & -Z \end{bmatrix}, Q\right) \\
&= Pe^{-hs}\mathcal{H}_l\left(\begin{bmatrix} M^{-1} & -\tilde{X}_0 - \tilde{Y}Z \\ 0 & \tilde{N}e^{-hs} \end{bmatrix}, Q\right) \\
&= Pe^{-hs}(M^{-1})^{-1}(\tilde{X}_0 + \tilde{Y}Z + Q\tilde{N}e^{-hs}) \\
&= N(\tilde{X}_0 + \tilde{Y}Z + Q\tilde{N}e^{-hs})e^{-hs}. \quad (9.13)
\end{aligned}$$

Since all the terms in the above formula are stable, the disturbance response is stable.

Obviously, the set-point response $G_{yr}(s)$ is only affected by the first degree-of-freedom $F(s)$ while the disturbance response $G_{yd}(s)$ is only affected by the second degree-of-freedom $Q(s)$. They are decoupled from each other. The fact that M does not appear in both transfer matrices means that it is possible to factorise the dominant poles and/or unstable poles into M. As a result, the disturbance response is no longer dominated by the process poles. See Subsection 2.3.3 for an example.

For stable processes, $P(s) \in H_\infty$, the double co-prime factorisation may be chosen as

9.2 Two-degree-of-freedom Realization of the Controller

$$\tilde{M} = M = I,$$
$$\tilde{N} = N = P(s), \quad \text{and} \quad \tilde{N}_0 = N_0 = P_0(s), \quad (9.14)$$
$$\tilde{X}_0 = X_0 = I \quad \text{and} \quad \tilde{Y} = Y = 0.$$

As a result, the transfer matrices are

$$G_{yr}(s) = P(s)F(s)e^{-hs},$$
$$G_{yd}(s) = P(s)\left(I + QP(s)e^{-hs}\right)e^{-hs}.$$

The disturbance response will be dominated by the process modes if the time constant is too large. A solution provided here is to factorise the slow modes into M and $\tilde{M}$ for stable processes to obtain better disturbance response. However, by doing this the robustness of the system is limited [4].

9.2.4 Robustness Analysis

The output loop transfer matrix of the control system shown in Figure 9.1 is

$$L(s) = P(s)e^{-hs}G_{uy}(s)$$
$$= P(s)e^{-hs}(I - \tilde{K}_0 Z)^{-1}\tilde{K}_0$$
$$= Ne^{-hs}\left(I + (-\tilde{Y} + Q\tilde{M})Ne^{-hs}\right)^{-1}(-\tilde{Y} + Q\tilde{M}).$$

Thus, the corresponding sensitivity function is

$$S(s) = (I - L(s))^{-1}$$
$$= I + Ne^{-hs}(-\tilde{Y} + Q\tilde{M})$$

and the complementary sensitivity function is

$$T(s) = I - S(s) = -Ne^{-hs}(-\tilde{Y} + Q\tilde{M}). \quad (9.15)$$

According to the well-known robust control theory [180], the closed-loop system, which is nominally stable, is robustly stable if $\left\|N(-\tilde{Y} + Q\tilde{M})\Delta(s)\right\|_\infty < 1$ for a multiplicative uncertainty $\Delta(s) \in H_\infty$. For stable processes $P(s) \in H_\infty$, the co-prime factorisation (9.14) holds and the system is robustly stable if $\|Q(s)\|_\infty < \frac{1}{\|P(s)\|_\infty \|\Delta(s)\|_\infty}$ for $\Delta(s) \in H_\infty$.

9.2.5 Ideal Disturbance Response

Introduce another delay-free transfer matrix $P_1(s)$ realized as

$$P_1(s) = \left[\begin{array}{c|c} A & B \\ \hline Ce^{Ah} & 0 \end{array}\right] = Ce^{Ah}(sI - A)^{-1}B \quad (9.16)$$

and the FIR block
$$Z_1(s) = P(s) - P_1(s)e^{-hs}.$$
Z_1 is stable because
$$Z_1(s) = C(sI - A)^{-1}B - Ce^{Ah}(sI - A)^{-1}Be^{-hs}$$
$$= C(I - e^{-(sI-A)h})(sI - A)^{-1}B$$
is also an entire function of s.

If F and L are chosen such that $A_F = A + BF$ and $A_L = A + LC$ are both stable (hence, $e^{Ah}(A + LC)e^{-Ah}$ is stable), then the factorisations (9.7) of $P(s)$ can be realized, according to the formulas in Subsection 3.3.1, as

$$\begin{bmatrix} M \\ N \end{bmatrix} = \left[\begin{array}{c|c} A+BF & B \\ \hline F & I \\ C & 0 \end{array} \right],$$

$$\begin{bmatrix} -\tilde{N} & \tilde{M} \end{bmatrix} = \left[\begin{array}{c|cc} A+LC & -B & L \\ \hline C & 0 & I \end{array} \right]$$

and the doubly co-prime factorisations (9.3) of $P_0(s)$ can be implemented [180] as

$$\begin{bmatrix} M & -Y \\ N_0 & X_0 \end{bmatrix} = \left[\begin{array}{c|cc} A+BF & B & -e^{Ah}L \\ \hline F & I & 0 \\ Ce^{-Ah} & 0 & I \end{array} \right],$$

$$\begin{bmatrix} \tilde{X}_0 & \tilde{Y} \\ -\tilde{N}_0 & \tilde{M} \end{bmatrix} = \left[\begin{array}{c|cc} e^{Ah}(A+LC)e^{-Ah} & -B & e^{Ah}L \\ \hline F & I & 0 \\ Ce^{-Ah} & 0 & I \end{array} \right].$$

The transfer matrix from disturbance $d(s)$ to output $Y(s)$ (9.13) can be further simplified to

$$G_{yd}(s) = \left(Z_1 + \left(P_1 - N\tilde{Y}P + NQ\tilde{N} \right) e^{-hs} \right) e^{-hs}. \tag{9.17}$$

As can be seen from the next subsection,

$$P_1 - N\tilde{Y}P = \left[\begin{array}{c|c} A+BF & e^{Ah}LC \\ \hline -C & Ce^{Ah} \end{array} \right] \left[\begin{array}{c|c} A+LC & B \\ \hline I & 0 \end{array} \right].$$

This is stable no matter whether $P(s)$ is stable or not. Hence, the disturbance response (9.17) is divided into two parts: one is the FIR part Z_1 which acts only on $t \in [0, h]$ (assume that the disturbance is applied at $t = -h$ hereafter) and remains unchanged when $t > h$, and the other is the infinite-impulse-response (IIR) part $P_1(s) - N\tilde{Y}P(s) + NQ\tilde{N}$ which acts only when $t > h$. In other words, the disturbance response is *decoupled in the time domain*. The effect of the FIR Z_1 on the (unit-step) disturbance response when $t > h$ is a constant given by

9.2 Two-degree-of-freedom Realization of the Controller

$$Z_1(0) = C(e^{Ah} - I)A^{-1}B.$$

If the free parameter $Q(s)$ is chosen to meet the following equation:

$$Z_1(0) + P_1 - N\tilde{Y}P + NQ\tilde{N} = 0, \tag{9.18}$$

then the disturbance response when $t > h$ is 0. This is the ideal case. If $N(s)$ and $\tilde{N}(s)$ are of minimum-phase and invertible, then the ideal $Q(s)$ can be obtained as

$$Q_{opt} = -N^{-1}\left(Z_1(0) + P_1 - N\tilde{Y}P\right)\tilde{N}^{-1}. \tag{9.19}$$

In general, this transfer matrix is improper because the ideal disturbance response is not obtainable. A low-pass filter should be introduced to make it proper. A possible way is to choose an appropriate n such that

$$Q(s) = \frac{1}{(\lambda s + 1)^n} Q_{opt}(s)$$

is proper. The resulting disturbance response is then subideal.

9.2.6 Realization of $P_1 - N\tilde{Y}P$

$$P_1(s) - N\tilde{Y}P(s)$$

$$= \left[\begin{array}{c|c} A & B \\ \hline Ce^{Ah} & 0 \end{array}\right] - \left[\begin{array}{c|c} A_F & B \\ \hline C & 0 \end{array}\right] \left[\begin{array}{c|c} e^{Ah}A_L e^{-Ah} & e^{Ah}L \\ \hline F & 0 \end{array}\right] \left[\begin{array}{c|c} A & B \\ \hline C & 0 \end{array}\right]$$

$$= C(e^{Ah} - \left[\begin{array}{c|c} A_F & B \\ \hline I & 0 \end{array}\right] \left[\begin{array}{c|c} A_L & LC \\ \hline Fe^{Ah} & 0 \end{array}\right]) \left[\begin{array}{c|c} A & B \\ \hline I & 0 \end{array}\right]$$

$$= C \left[\begin{array}{ccc|c} A_F & BFe^{Ah} & 0 \\ 0 & A_L & LC \\ -I & 0 & e^{Ah} \end{array}\right] \left[\begin{array}{c|c} A & B \\ \hline I & 0 \end{array}\right]$$

$$= C \left[\begin{array}{cccc|c} A_F & BFe^{Ah} & 0 & 0 \\ 0 & A_L & LC & 0 \\ 0 & 0 & A & B \\ -I & 0 & e^{Ah} & 0 \end{array}\right]$$

$$= C \left[\begin{array}{cccc|c} A+BF & BF & 0 & 0 \\ 0 & A+e^{Ah}LCe^{-Ah} & e^{Ah}LCe^{-Ah} & 0 \\ 0 & 0 & A & e^{Ah}B \\ -I & 0 & I & 0 \end{array}\right]$$

$$= C \left[\begin{array}{cccc|c} A+BF & BF & BF & -e^{Ah}B \\ 0 & A+e^{Ah}LCe^{-Ah} & e^{Ah}LCe^{-Ah} & 0 \\ 0 & 0 & A & e^{Ah}B \\ -I & 0 & 0 & 0 \end{array}\right]$$

$$= C \left[\begin{array}{ccc|c} A+BF & BF & 0 & -e^{Ah}B \\ 0 & A+e^{Ah}LCe^{-Ah} & 0 & e^{Ah}B \\ 0 & 0 & A & e^{Ah}B \\ \hline -I & 0 & 0 & 0 \end{array}\right]$$

$$= C \left[\begin{array}{cc|c} A+e^{-Ah}BFe^{Ah} & e^{-Ah}BFe^{Ah} & -B \\ 0 & A+LC & B \\ \hline -e^{Ah} & 0 & 0 \end{array}\right]$$

$$= C \left[\begin{array}{cc|c} A+e^{-Ah}BFe^{Ah} & LC & 0 \\ 0 & A+LC & B \\ \hline -e^{Ah} & e^{Ah} & 0 \end{array}\right]$$

$$= \left[\begin{array}{c|c} A+e^{-Ah}BFe^{Ah} & LC \\ \hline -Ce^{Ah} & Ce^{Ah} \end{array}\right] \left[\begin{array}{c|c} A+LC & B \\ \hline I & 0 \end{array}\right]$$

$$= \left[\begin{array}{c|c} A+BF & e^{Ah}LC \\ \hline -C & Ce^{Ah} \end{array}\right] \left[\begin{array}{c|c} A+LC & B \\ \hline I & 0 \end{array}\right].$$

9.3 Application to Integral Processes with Dead Time

The control problem of processes with an integrator and long dead time has received much attention in recent years [8, 72, 159, 176, 178]. Since the process is not stable, the classical Smith predictor cannot be used. The approach discussed in the previous sections are applied to such systems here as an example[3]. An integral process with dead time can be expressed as

$$G_p(s) = \frac{k}{s} e^{-hs}, \qquad (9.20)$$

where k is the velocity gain and h is the delay. Here,

$$P(s) = P_0(s) = P_1(s) = \frac{k}{s} = \left[\begin{array}{c|c} 0 & k \\ \hline 1 & 0 \end{array}\right].$$

They can be factorised as (for simplicity, $M = \tilde{M}$ hereafter)

$$P(s) = P_0(s) = P_1(s) = \frac{k}{s+a} \cdot \left(\frac{s}{s+a}\right)^{-1}$$

with $a > 0$. The predictors are:

$$Z(s) = P_0(s) - P(s)e^{-hs} = \frac{1-e^{-hs}}{s} k,$$

---

[3] Portions reprinted, with permission, from [93, 94]. ©IEEE.

9.3 Application to Integral Processes with Dead Time

$$Z_1(s) = P(s) - P_1(s)e^{-hs} = \frac{1-e^{-hs}}{s}k.$$

Since the values at the only pole are

$$Z(0) = Z_1(0) = \lim_{s \to 0} \frac{1-e^{-hs}}{s}k = kh \neq \infty,$$

the predictors Z and Z_1 are stable.

The double co-prime factorisations of $P_0(s)$ are

$$\begin{bmatrix} \tilde{X}_0 & \tilde{Y} \\ -\tilde{N}_0 & \tilde{M} \end{bmatrix} = \begin{bmatrix} -a & -k & -a \\ -\frac{a}{k} & 1 & 0 \\ 1 & 0 & 1 \end{bmatrix} = \begin{bmatrix} \frac{s+2a}{s+a} & \frac{1}{k}\frac{a^2}{s+a} \\ -\frac{k}{s+a} & \frac{s}{s+a} \end{bmatrix},$$

$$\begin{bmatrix} M & -Y \\ N_0 & X_0 \end{bmatrix} = \begin{bmatrix} -a & k & a \\ -\frac{a}{k} & 1 & 0 \\ 1 & 0 & 1 \end{bmatrix} = \begin{bmatrix} \frac{s}{s+a} & -\frac{1}{k}\frac{a^2}{s+a} \\ \frac{k}{s+a} & \frac{s+2a}{s+a} \end{bmatrix}.$$

The controller is then parameterised as

$$K(s) = \left(1 - \tilde{K}_0(s)Z\right)^{-1} \tilde{K}_0(s) \tag{9.21}$$

with the main controller

$$\tilde{K}_0(s) = \left(\frac{s+2a}{s+a} + \frac{k}{s+a}Q\right)^{-1}\left(-\frac{1}{k}\frac{a^2}{s+a} + \frac{s}{s+a}Q\right). \tag{9.22}$$

If it is realized as the structure shown in Figure 9.1, then the transfer functions from the reference signal $R(s)$, the disturbance $d(s)$ to the output $Y(s)$ for the nominal case are, respectively,

$$G_{yr}(s) = \frac{k}{s+a}F(s)e^{-hs},$$

$$G_{yd}(s) = \left(\frac{1-e^{-hs}}{s}k + k\left(\frac{(s+2a)}{(s+a)^2} + \frac{k}{(s+a)^2}Q\right)e^{-hs}\right)e^{-hs}. \tag{9.23}$$

In order to obtain zero static error under the step change from the set-point and the disturbance, it is necessary that

$$F(0) = \frac{a}{k} \quad \text{and} \quad Q(0) = -\frac{a}{k}(2+ah).$$

The corresponding ideal $Q_{opt}(s)$ from (9.19) is

$$Q_{opt}(s) = -N^{-1}\left(Z_1(0) + P_1(s) - N\tilde{Y}P(s)\right)\tilde{N}^{-1}$$

$$= -\frac{kh + \frac{k}{s} - \frac{k}{s+a} \cdot \frac{1}{k}\frac{a^2}{s+a} \cdot \frac{k}{s}}{\left(\frac{k}{s+a}\right)^2}$$

$$= -\frac{1}{k}\left(h(s+a)^2 + s + 2a\right).$$

Q_{opt} is not proper and the ideal disturbance response is not obtainable. One way to obtain a subideal disturbance response is to make $Q_{opt}(s)$ proper as

$$Q(s) = -\frac{1}{k}\left(\frac{h(s+a)^2}{(\lambda s+1)^2} + \frac{\lambda(2-\lambda a)s+1}{(\lambda s+1)^2}(s+a) + a\right),$$

where $\lambda > 0$ is a tuning parameter. If λ is small enough, then the disturbance response is fast enough. As shown later, such a way of making $Q(s)$ proper ensures that neither the disturbance response nor the robust stability is affected by the free parameter a. The resulting main controller $\tilde{K}_0(s)$ (9.22), the controller $K(s)$ (9.21), disturbance response (9.23) and complementary sensitivity function are listed below:

$$\tilde{K}_0(s) = \left(\frac{(\lambda^2 s - h)(s+a)}{(\lambda s+1)^2}\right)^{-1}\left(-\frac{1}{k}\frac{((2\lambda+h)s+1)(s+a)}{(\lambda s+1)^2}\right),$$

$$K(s) = -\frac{1}{k}\frac{((2\lambda+h)s+1)s}{(\lambda s+1)^2 - ((2\lambda+h)s+1)e^{-hs}}, \quad (9.24)$$

$$G_{yd}(s) = k\left(\frac{1-e^{-hs}}{s} + \frac{\lambda^2 s - h}{(\lambda s+1)^2}e^{-hs}\right)e^{-hs}$$

$$= \frac{k}{s}\left(1 - \frac{(2\lambda+h)s+1}{(\lambda s+1)^2}e^{-hs}\right)e^{-hs}, \quad (9.25)$$

$$T(s) = -Ne^{-hs}(-\tilde{Y} + Q\tilde{M}) = \frac{(2\lambda+h)s+1}{(\lambda s+1)^2}e^{-hs}.$$

The disturbance response (9.25) has been obtained by several different schemes [107, 156, 174, 177]. The achievable specifications and robustly stable region can be found in [175, 176].

The rest of this section is devoted to analysing the stability of the controller. The characteristic equation of the controller (9.24) is equivalent to

$$\frac{(\lambda s+1)^2}{(2\lambda+h)s+1} = e^{-hs}.$$

The stability can be judged by using the dual-locus [127, 164] shown in Figure 9.2. If the two loci do not intersect, then the controller is stable. Otherwise, it is unstable. In other words, if the locus of $\frac{(\lambda s+1)^2}{(2\lambda+h)s+1}$ crosses the unit circle before the e^{-hs} locus arrives, then the system is stable; if e^{-hs} arrives earlier than $\frac{(\lambda s+1)^2}{(2\lambda+h)s+1}$ then the controller is unstable.

Denote $\beta = \frac{\lambda}{h}$, the locus of $\frac{(\lambda s+1)^2}{(2\lambda+h)s+1}$ intersects with the unit circle at

$$\omega_c = \frac{\sqrt{2\beta^2 + 4\beta + 1}}{\beta^2 h}.$$

At this frequency, the phase of $\frac{(\lambda s+1)^2}{(2\lambda+h)s+1}$ is

$$\phi_\lambda = 2\arctan\frac{\sqrt{2\beta^2+4\beta+1}}{\beta} - \arctan\frac{(2\beta+1)\sqrt{2\beta^2+4\beta+1}}{\beta^2}.$$

If the locus of e^{-hs} also arrives at this point at ω_c, then the following equation should be met:

$$\phi_\lambda = 2\pi - h\omega_c.$$

This equation has a unique solution at about $\beta = 0.63$. For $\beta > 0.63$, the locus of $\frac{(\lambda s+1)^2}{(2\lambda+h)s+1}$ crosses the unit circle before the e^{-hs} locus arrives. As a result, the controller (9.24) is stable for $\beta > 0.63$.

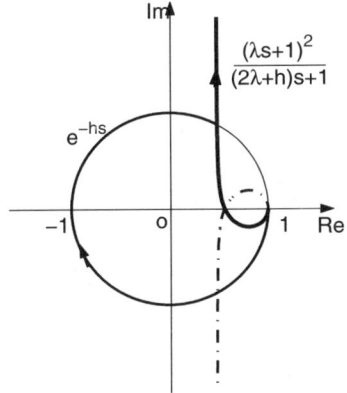

Figure 9.2. The dual locus to judge controller stability

9.4 Summary

This chapter presents the co-prime factorisation of all stabilising controllers for stable and unstable processes with dead time. The controller incorporates a modified Smith predictor, which is an entire function of s and belongs to H_∞. With a realization having two degrees-of-freedom, the disturbance response and the set-point response are decoupled and can be designed separately. The realization in the chain-scattering representation gives clear insight into the two degrees-of-freedom and the differences between the controllers for processes with and without dead time. The constraint required on free parameter $Q(s)$ to obtain the ideal disturbance response is obtained. As an example, the approach is applied to integral processes with dead time.

10

Unified Smith Predictor

Modified Smith predictors (MSP) play a very important role in the control of time-delay systems, as can be seen in previous chapters. In this chapter,[1] a numerical problem associated with the MSP is identified and an alternative predictor, named the *unified Smith predictor* (USP), is proposed to overcome this problem. The proposed USP combines the classical Smith predictor with the modified one, after spectral decomposition of the plant. An equivalent representation of the original delay system, together with the USP, is derived. Based on this representation, all the stabilising controllers are parameterised and the standard H^2 problem and a transformed H^∞ problem are solved.

10.1 Introduction

The classical *Smith predictor* (SP) [126, 127] is an effective tool to reduce control problems (such as pole assignment or tracking) for a finite-dimensional LTI stable system with an input or output delay to corresponding delay-free problems. A finite-spectrum assignment scheme was developed in [70, 108] to handle input delays in *unstable* plants by state feedback, using the predicted state of the delay system. Watanabe and Ito [147] overcame some shortcomings in finite-spectrum assignment and other process-model-based control schemes available then by using a Smith-predictor-like block, which was afterwards called a *modified Smith predictor* (MSP); see [113] and references therein. As shown in Section 2.5 and in [91], the modified Smith predictor and finite-spectrum assignment are actually equivalent for systems with a single I/O delay. Recently, prediction has been recognised as a fundamental concept for the stabilisation of delay systems [89, 91]. Similar predictor blocks have played an important role in H^∞ control of time-delay systems [26, 80, 87, 160, 161, 162] and in continuous-time deadbeat control [106, 159].

---

[1] Portions reprinted, with permission, from [179]. ©Taylor & Francis.

The modified Smith predictor may run into numerical problems for delay systems with fast stable eigenvalues. Indeed, the matrix exponential e^{-Ah} (where h is the delay) appearing in the MSP may be practically non-computable for such systems. Such a numerical problem was mentioned in [80, p. 279] and a technique, which is not systematic, was suggested in [183] to overcome the problem. This problem was attributed in these papers to large delays. In fact, this numerical problem might occur even for very small delays (with respect to the time constant of the system) if there are some stable eigenvalues λ and they are very fast with respect to the delay h (*i.e.*, the product $|\text{Re}\lambda|\,h$ is large). An alternative predictor, called the *unified Smith predictor* (USP), is proposed to overcome this problem. The USP combines the features of the SP and the MSP and does not require computation of the matrix exponential for the stable eigenvalues. This is achieved by decomposing the state space of the finite-dimensional part of the plant into unstable and stable invariant subspaces. The controller design techniques based on the MSP have to be re-considered for the USP, to make them applicable in practice. For this reason, an equivalent representation for the *augmented plant*, which consists of the original plant together with the USP, is proposed. This equivalent representation is then used to give a stabilising controller parameterisation and to solve the standard H^2 problem and a transformed H^∞ problem. Further research is needed to solve the H^∞ control problem for a delay system with a USP.

This chapter is organised as follows. The numerical problem with the MSP is identified in Section 10.3 using a very simple example, and then the USP is introduced. An equivalent representation of the augmented plant is derived in Section 10.4 and then it is shown in Section 10.5 how this equivalent representation can be used to derive a stabilising controller parameterisation, to solve the H^2 problem and to solve a transformed H^∞ problem.

10.2 Predictor-based Control Structure

As before, this chapter is written from a frequency domain perspective, which means that LTI systems are represented by their transfer functions. Here, a transfer function is an analytic function defined on a domain which contains a half-plane $\mathbb{C}_\alpha = \{s \in \mathbb{C} \mid \text{Re}\, s > \alpha\}$. A transfer function is called *exponentially stable* if it is bounded on a half-plane $\mathbb{C}_\alpha$ with $\alpha < 0$. Obviously, exponentially stable transfer functions are contained in H^∞.

Consider a dead-time plant P_h, with $P_h(s) = P(s)e^{-sh}$, where $h > 0$ and the rational part P is realized as

$$P = \left[\begin{array}{c|c} A & B \\ \hline C & D \end{array}\right]. \tag{10.1}$$

A *predictor-based controller* for the dead-time plant P_h consists of a predictor Z and a *stabilising compensator* C, as shown in Figure 10.1 (see also Figure

10.5 for a more general structure). A *predictor* for P_h is an exponentially stable system Z such that the *augmented plant* $P^{\text{aug}} = P_h + Z$ is rational and $v + y$ (see Figure 10.1) is a predicted signal of y. The underlying idea is the well-known fact that there is a one-to-one correspondence between the set of all the stabilising controllers for P_h and for P^{aug} (when Z is stable); see for example Remark 3.6 and Example 4.1 in [19].

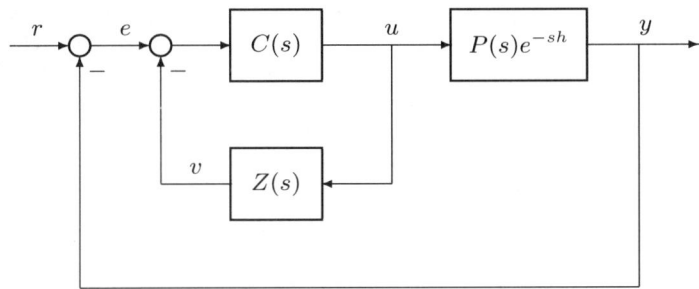

Figure 10.1. Dead-time plant with a predictor-based controller

In order to simplify the exposition, only the stabilisation problem will be studied in this section. If the tracking problem is considered, then an additional constraint on C is required. For example, if the reference r is a step signal then it is required that $\lim_{s \to 0} C(s)(I + Z(s)C(s))^{-1} = \infty$ [113, 147].

If P is stable, then the predictor Z can be chosen to be the *classical Smith predictor* (SP) [126, 127],

$$Z_{\text{SP}}(s) = P(s) - P(s)e^{-sh}, \qquad (10.2)$$

and the stabilising compensator C is designed as a stabilising controller for the delay-free system P (in this case, $P^{\text{aug}} = P$). If P is unstable, then the predictor Z can be chosen to be the *modified Smith predictor* (MSP) [113, 147],

$$Z_{\text{MSP}}(s) = P^{\text{aug}}(s) - P(s)e^{-sh}, \qquad (10.3)$$

where $P^{\text{aug}} = P_h + Z_{\text{MSP}}$ is the *augmented plant* given by

$$P^{\text{aug}} = \left[\begin{array}{c|c} A & B \\ \hline Ce^{-Ah} & 0 \end{array} \right] = \left[\begin{array}{c|c} A & e^{-Ah}B \\ \hline C & 0 \end{array} \right]. \qquad (10.4)$$

Note that Z_{MSP} has its impulse response supported on $[0, h]$, hence it is exponentially stable. Now the stabilising compensator C is designed as a stabilising controller for P^{aug}. Further background and applications of the MSP can be found in [80, 91, 113]. See also Chapter 2. Implementing Z_{MSP} is a delicate problem, as can be seen in Part II, because the hidden unstable modes have to be removed to guarantee the internal stability.

The condition that Z is stable means that all the (unstable) poles of P have to be included in P^{aug}. Hence, a stabilising controller C for P^{aug} also stabilises P. The transfer function from u to y is

$$T_{yu}(s) = e^{-sh} P(I + CP^{\mathrm{aug}})^{-1} + e^{-sh} P(I + CP^{\mathrm{aug}})^{-1} C \cdot Z.$$

It is clear that the (stable or unstable) poles of P which are included in P^{aug} can be shifted, if these poles do not appear in Z. In other words, whether an open-loop pole appears in the closed-loop input-disturbance response is determined by Z, when the controller C is properly designed. Hence, it is possible to remove open-loop poles from the closed-loop input-disturbance response for a stable plant, by choosing a suitable Z which does not include the open-loop poles (e.g., the MSP).

10.3 Problem Identification and the Solution

10.3.1 A Numerical Problem with the MSP

Now consider a simple (but somewhat extreme) example with

$$P = \left[\begin{array}{cc|c} -1000 & 0 & 1 \\ 0 & 1 & 1 \\ \hline 1 & 1 & 0 \end{array}\right],$$

i.e., $P(s) = \frac{1}{s+1000} + \frac{1}{s-1}$. According to (10.3), the predictor needed is

$$Z_{\mathrm{MSP}}(s) = \frac{e^{1000h} - e^{-sh}}{s + 1000} + \frac{e^{-h} - e^{-sh}}{s - 1}.$$

Clearly, there is a numerical problem: e^{1000h} is a huge number even for a not so large delay h! Indeed, according to IEEE Standard 754 [50], which specifies today the most common representation for real numbers on computers, including Intel-based PC's, Macintoshes and most Unix platforms, this number is considered to be $+\infty$ (INF) for $h \geq 0.71$ s. Certainly, such a component in a controller is not allowed in practice.

This problem arises due to a fast stable pole (here, $p = -1000$) of P. A stable pole makes the real part of the exponent positive, and if this is large, the numerical problem occurs. If the plant is completely unstable (there are no stable poles), then there will be no such problem. Thus, if the plant has fast stable poles, then the predictor Z_{MSP} from (10.3) should not be used.

10.3.2 The Unified Smith Predictor (USP)

A natural solution to the numerical problem encountered in the previous subsection is to decompose P into the sum of a stable part P_s and an unstable part P_u and then to construct predictors for them, using the classical Smith predictor (10.2) for $P_s(s)e^{-sh}$ and the modified Smith predictor (10.3) for $P_u(s)e^{-sh}$. A new term, *unified Smith predictor* (USP), is proposed for such a predictor.

As is well known, the rational part of the plant, P given in (10.1), can be split (decomposed) into the sum of a stable part P_s and an unstable part P_u, e.g., by applying a suitable linear coordinate transformation in its state space. There exist a lot of such similarity transformations. One of them, denoted here by V, can be obtained by bringing the system matrix A to the Jordan canonical form $J_D = V^{-1}AV$. Assume that V is a nonsingular matrix such that

$$P = \left[\begin{array}{c|c} V^{-1}AV & V^{-1}B \\ \hline CV & D \end{array}\right] = \left[\begin{array}{cc|c} A_u & 0 & B_u \\ 0 & A_s & B_s \\ \hline C_u & C_s & D \end{array}\right], \tag{10.5}$$

where A_s is stable and A_u is completely unstable, then P can be split as $P = P_s + P_u$ with

$$P_s = \left[\begin{array}{c|c} A_s & B_s \\ \hline C_s & 0 \end{array}\right] \quad \text{and} \quad P_u = \left[\begin{array}{c|c} A_u & B_u \\ \hline C_u & D \end{array}\right].$$

As a matter of fact, the decomposition can be done by splitting the complex plane along any vertical line $\text{Re}\, s = \alpha$ with $\alpha \leq 0$. Then the eigenvalues of A_u are all the eigenvalues λ of A with $\text{Re}\,\lambda \geq \alpha$, while A_s has the remaining eigenvalues of A. In the sequel, in order to simplify the exposition, the complex plane is split along the imaginary axis ($\alpha = 0$).

The predictor for the stable part P_s can be taken as a classical SP,

$$Z_s(s) = P_s(s) - P_s(s)e^{-sh}, \tag{10.6}$$

and the predictor for the unstable part P_u can be taken as the following MSP:

$$Z_u(s) = P_u^{\text{aug}}(s) - P_u(s)e^{-sh}, \quad P_u^{\text{aug}} \doteq \left[\begin{array}{c|c} A_u & B_u \\ \hline C_u e^{-A_u h} & 0 \end{array}\right].$$

The USP for the plant $P_h(s) = P(s)e^{-sh}$ defined by $Z = Z_s + Z_u$ is shown in Figure 10.2. It is now clear that the USP is

$$Z(s) = P^{\text{aug}}(s) - P(s)e^{-sh}, \tag{10.7}$$

where $P^{\text{aug}} = P_s + P_u^{\text{aug}}$ and a realization of P^{aug} is

$$P^{\text{aug}} = \left[\begin{array}{c|c} A & B \\ \hline CE_h & 0 \end{array}\right] = \left[\begin{array}{c|c} A & E_h B \\ \hline C & 0 \end{array}\right], \quad E_h = V \left[\begin{array}{cc} e^{-A_u h} & 0 \\ 0 & I_s \end{array}\right] V^{-1}. \tag{10.8}$$

Here, the identity I_s has the same dimension as A_s and E_h is commutable with A. The impulse response of the USP is not finite, unlike that of the MSP.

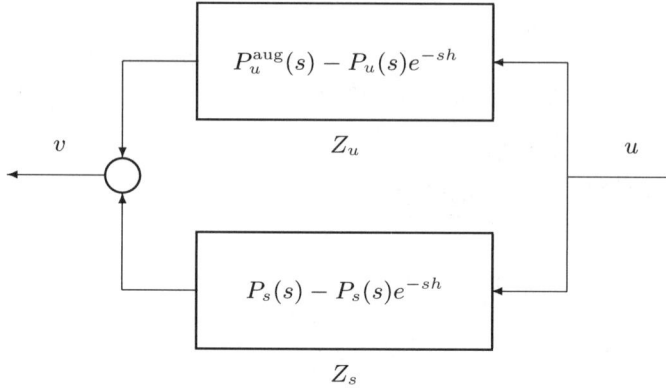

Figure 10.2. Unified Smith predictor $Z = Z_s + Z_u$

The above P^{aug} is the *augmented plant*, obtained by connecting the original plant and the USP in parallel. The stabilising compensator C in Figure 10.1 should be designed as a stabilising controller for P^{aug}.

The use of the different predictors (SP, MSP and USP) is summarised in Table 10.1: when the plant is stable, the USP reduces to the SP; when the plant is completely unstable, the USP reduces to the MSP; when the plant is unstable but has some stable poles (mixed), then the USP should be used, especially if some stable poles are fast (as explained earlier).

Table 10.1. USP needed for different types of plants

Type of plant	stable	completely unstable	mixed
Type of USP	SP	MSP	USP

Returning to the example in the previous subsection, the USP for it is given by
$$Z(s) = \frac{1 - e^{-sh}}{s + 1000} + \frac{e^{-h} - e^{-sh}}{s - 1}.$$

Remark 10.1. A possible realization of Z consists of the finite-dimensional system
$$\dot{z}(t) = Az(t) + E_h Bu(t) - Bu(t - h)$$
$$v(t) = Cz(t) - Du(t - h)$$

together with a realization of a delay line (which is needed to generate the signal $u(t - h)$). However, such a realization of Z would have hidden unstable modes. An alternative is to realize Z_s and Z_u separately. The first is not a problem and the second can be implemented using the techniques discussed in Part II.

10.3 Problem Identification and the Solution

Remark 10.2. Z_s defined in (10.6) could be replaced by

$$Z_s(s) = -P_s(s)e^{-sh},$$

as in internal model control [99]. In this case, P^{aug} in (10.8) would change, since in the definition of E_h the identity I_s would have to be replaced by zero. Now P^{aug} would be simpler (it would have a lower order). This may be advantageous in the context of Figure 10.1. However, in the more general framework of Figure 10.5 (in the next section) this may cause a rank problem.

Remark 10.3. The USP (in particular, the SP and the MSP) can be generalised to multiple delays, *i.e.*, to the situation where the components u_1, u_2, ... u_m of the vector u are delayed by $h_1, h_2, \ldots h_m \geq 0$. Denote $H = \text{diag}(h_1, h_2, \ldots h_m)$ and

$$E \otimes B = \begin{bmatrix} E_{h_1} b_1 & E_{h_2} b_2 & \ldots & E_{h_m} b_m \end{bmatrix},$$

where b_k is the k-th column of B, *i.e.*, $B = \begin{bmatrix} b_1 & b_2 & \ldots & b_m \end{bmatrix}$. Then

$$P^{\text{aug}} = \left[\begin{array}{c|c} A & E \otimes B \\ \hline C & 0 \end{array}\right], \qquad Z(s) = P^{\text{aug}}(s) - P(s)e^{-Hs} \qquad (10.9)$$

and Z is exponentially stable.

Remark 10.4. An implementation of Z from Remark 10.3 (the USP for multiple delays) which avoids the problem of unstable modes is outlined here. The components Z_u and Z_s are separately implemented, but using the same delay lines for both, as shown in Figure 10.3. The component Z_u is a hybrid system, containing two copies of an LTI block with the possibility of resetting, and the switches S_a, S_b and S_v. In this diagram, denote $B_u = \begin{bmatrix} \beta_1 & \beta_2 & \ldots & \beta_m \end{bmatrix}$ and $E \otimes B_u = \begin{bmatrix} e^{-A_u h_1} \beta_1 & e^{-A_u h_2} \beta_2 & \ldots & e^{-A_u h_m} \beta_m \end{bmatrix}$. The output of the USP is $v = v^s + v^u$, where $v^s = Z_s u$ and $v^u = Z_u u$. v^u is produced by one of the two LTI blocks which can be reset by the signals R_a and R_b, respectively. When resetting one of them, say the top one, at a time τ, its second input u^a is switched from u^d to 0 using the switch S_a. Afterwards, u_k^a is reconnected to u_k^d at the time $\tau + h_k$ ($k = 1, 2, \ldots m$). During this time, $v^u = v^b$. After all the components of u^d have been reconnected to the upper LTI block, so that $u^a = u^d$, its output v^a will be the desired value of v^u, and so the switch S_v can be set so that $v^u = v^a$. The switches could, theoretically, remain in this position forever, but since A_u is unstable, tiny computational or rounding errors or the effect of noise will grow and will corrupt the output v^a and hence v^u. To prevent this, the other LTI block is reset (while switching the components of u^b, similarly as above, using the switch S_b) and, when the output v^b becomes correct, S_v is switched so that $v^u = v^b$. This cycle repeats itself indefinitely and it can be shown that the USP is stable and produces the correct output. This implementation is related to the periodic resetting mechanism for integrators encountered in [135] and the resetting Smith predictor in [96].

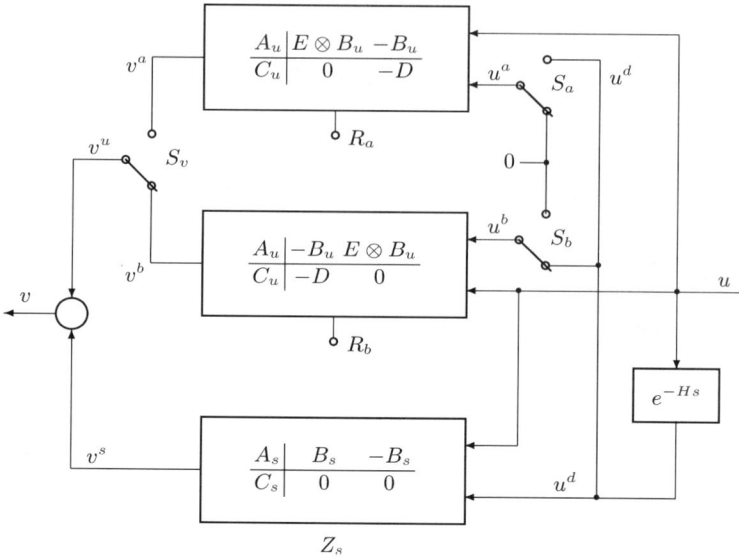

Figure 10.3. Implementation of the USP (10.9) for multiple delays, using two resetting LTI systems to implement Z_u. The logical controller which controls the switches and generates the resetting signals R_a and R_b (in an open-loop manner) is not shown.

10.4 Control Systems with a USP: Equivalent Diagrams

In this section a more general type of plant, with two inputs and two outputs (each of these signals may be vector-valued), is considered. The input signal w contains references and disturbances, u is the control input, z is the tracking error and y is the measurement available to the controller. Assume that the plant P_h consists of a rational part P and a delay by $h > 0$ acting on u, as shown in Figure 10.4 and denote

$$P = \begin{bmatrix} P_{11} & P_{12} \\ P_{21} & P_{22} \end{bmatrix} = \left[\begin{array}{c|cc} A & B_1 & B_2 \\ \hline C_1 & D_{11} & D_{12} \\ C_2 & D_{21} & D_{22} \end{array}\right]. \qquad (10.10)$$

K is called a *stabilising controller* for P_h if the transfer functions from the three external inputs shown in Figure 10.4 to any other signal in the diagram are in H^∞. K is called an *exponentially stabilising controller* for P_h if the same transfer functions are exponentially stable. In the sequel, the two external signals appearing in the lower part of Figure 10.4 will be taken to be zero.

As in the previous section, let V be a nonsingular matrix such that

$$V^{-1}AV = \begin{bmatrix} A_u & 0 \\ 0 & A_s \end{bmatrix},$$

10.4 Control Systems with a USP: Equivalent Diagrams 161

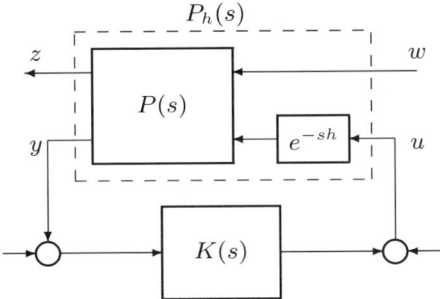

Figure 10.4. Control system comprising a dead-time plant P_h (with a rational part P) and a stabilising controller K

where A_u is completely unstable and A_s is stable. As in (10.7), the following USP designed for the component P_{22} of the plant is introduced:

$$Z(s) \doteq P_{22}^{\mathrm{aug}}(s) - P_{22}(s)e^{-sh}, \tag{10.11}$$

where, as in (10.8),

$$P_{22}^{\mathrm{aug}} \doteq \left[\begin{array}{c|c} A & B_2 \\ \hline C_2 E_h & 0 \end{array}\right], \qquad E_h = V\begin{bmatrix} e^{-A_u h} & 0 \\ 0 & I_s \end{bmatrix} V^{-1}.$$

By connecting this USP in parallel with the u to y component of P_h, as shown in Figure 10.5, thus creating a new measurement output y_p, a new *augmented plant*

$$P^{\mathrm{aug}}(s) = \begin{bmatrix} P_{11}(s) & P_{12}(s)e^{-hs} \\ P_{21}(s) & P_{22}^{\mathrm{aug}}(s) \end{bmatrix} \tag{10.12}$$

is obtained. It is well known (and easy to see from Figure 10.5) that C is a stabilising controller for P^{aug} if and only if $K = C(I - ZC)^{-1}$ is a stabilising controller for the original dead-time plant. The same statement remains true with 'exponentially stabilising' in place of 'stabilising' (and, of course, the corresponding set of controllers is smaller).

An equivalent representation of P^{aug} given below is useful for the problems treated in the next section. Proposition 10.1 and Remark 10.5 below are related to Lemma 1 in [91] (the corresponding notation is $\Delta_1 = Z_1$ and $\Delta_2 = Z$).

Proposition 10.1. *The augmented plant can be decomposed as*

$$P^{\mathrm{aug}}(s) = \begin{bmatrix} Z_1(s) & 0 \\ 0 & 0 \end{bmatrix} + \begin{bmatrix} e^{-sh}I & 0 \\ 0 & I \end{bmatrix} \tilde{P}(s), \tag{10.13}$$

where

$$Z_1(s) \doteq P_{11}(s) - \left[\begin{array}{c|c} A & e^{Ah}B_1 \\ \hline C_1 & 0 \end{array}\right](s)e^{-sh},$$

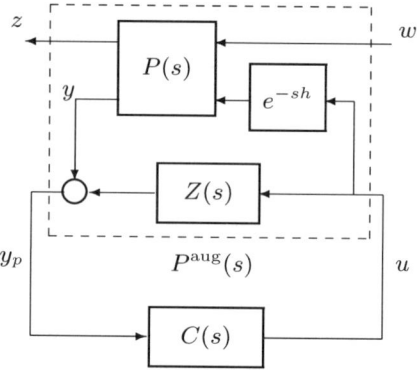

Figure 10.5. Control system from Figure 10.4, in which the controller K has been decomposed into a USP denoted Z and a stabilising compensator C, so that $K = C(I - ZC)^{-1}$.

$$\tilde{P} = \tilde{P}_0 + \begin{bmatrix} \tilde{P}_s & 0 \\ 0 & 0 \end{bmatrix}, \quad \tilde{P}_0 = \left[\begin{array}{c|cc} A & E_h^{-1} B_1 & B_2 \\ \hline C_1 & 0 & D_{12} \\ C_2 E_h & D_{21} & 0 \end{array} \right], \tag{10.14}$$

$$\tilde{P}_s = \left[\begin{array}{c|c} A_s & \left[0 \; e^{A_s h} - I_s \right] V^{-1} B_1 \\ \hline C_1 V \begin{bmatrix} 0 \\ I_s \end{bmatrix} & 0 \end{array} \right]. \tag{10.15}$$

The block diagram corresponding to the decomposition (10.13) and (10.14) of P^{aug} is shown in Figure 10.6.

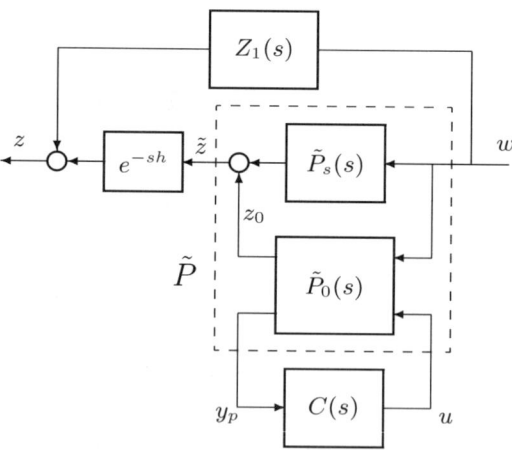

Figure 10.6. Equivalent representation of the control system in Figure 10.5, using the decomposition of P^{aug} given in Proposition 10.1.

10.4 Control Systems with a USP: Equivalent Diagrams

Proof. Using the partition (10.12) of P^{aug}, the formula (10.13) to be proven is equivalent to the following four formulas:

$$P_{11}(s) = Z_1(s) + \left(\left[\begin{array}{c|c} A & E_h^{-1} B_1 \\ \hline C_1 & 0 \end{array} \right] + \tilde{P}_s(s) \right) e^{-sh}, \qquad (10.16)$$

$$P_{12} e^{-sh} = \left[\begin{array}{c|c} A & B_2 \\ \hline C_1 & D_{12} \end{array} \right] e^{-sh},$$

$$P_{21} = \left[\begin{array}{c|c} A & E_h^{-1} B_1 \\ \hline C_2 E_h & D_{21} \end{array} \right], \qquad P_{22}^{\text{aug}} = \left[\begin{array}{c|c} A & B_2 \\ \hline C_2 E_h & 0 \end{array} \right].$$

The last three formulas are clearly true, so only the first formula (10.16) requires a little work. Since

$$\left[\begin{array}{c|c} A & E_h^{-1} B_1 \\ \hline C_1 & 0 \end{array} \right] = \left[\begin{array}{cc|c} A_u & 0 & \left[\begin{array}{cc} e^{A_u h} & 0 \\ 0 & I_s \end{array} \right] V^{-1} B_1 \\ 0 & A_s & \\ \hline C_1 V & & 0 \end{array} \right]$$

and

$$\tilde{P}_s = \left[\begin{array}{cc|c} A_u & 0 & \left[\begin{array}{cc} 0 & 0 \\ 0 & e^{A_s h} - I_s \end{array} \right] V^{-1} B_1 \\ 0 & A_s & \\ \hline C_1 V & & 0 \end{array} \right],$$

(10.16) follows easily. □

Note that if the USP would be chosen as described in Remark 10.2, then E_h would become singular, which is the rank problem mentioned in Remark 10.2, and hence the decomposition (10.14) of $\tilde{P}$ would have to be replaced by a more complicated one.

Remark 10.5. It follows from Proposition 10.1 that the closed-loop transfer function T_{zw} from w to z can be written as the sum of three terms:

$$T_{zw}(s) = Z_1(s) + \tilde{P}_s(s) e^{-sh} + \mathcal{F}_l(\tilde{P}_0(s), C(s)) e^{-sh}, \qquad (10.17)$$

where $\mathcal{F}_l(\tilde{P}_0, C)$ is the transfer function from w to z_0 in Figure 10.6. Z_1 is an FIR system with impulse response supported on $[0, h]$, while the second and third term have impulse responses supported on $[h, \infty)$. Hence, if these terms are in $H^2(\mathbb{C}_0)$, then the first term is orthogonal to the second and third term.

Remark 10.6. A realization of $\tilde{P}$ is given by

$$\tilde{P} = \left[\begin{array}{cc|ccc} A & 0 & & E_h^{-1} B_1 & B_2 \\ 0 & A_s & \left[0 \ e^{A_s h} - I_s \right] V^{-1} B_1 & 0 \\ \hline C_1 & C_1 V \left[\begin{array}{c} 0 \\ I_s \end{array} \right] & 0 & D_{12} \\ C_2 E_h & 0 & D_{21} & 0 \end{array} \right]. \qquad (10.18)$$

If P is completely unstable, then the dimensions of A_s and I_s are zero, and the block $\tilde{P}_s$ in Figure 10.6 disappears. In this case, Z reduces to an MSP and $\tilde{P}$ becomes

$$\tilde{P} = \left[\begin{array}{c|cc} A & e^{Ah}B_1 & B_2 \\ \hline C_1 & 0 & D_{12} \\ C_2 e^{-Ah} & D_{21} & 0 \end{array}\right] = \tilde{P}_0.$$

If P is stable, then $\tilde{P}$ is reduced to

$$\tilde{P} = \left[\begin{array}{cc|cc} A & 0 & B_1 & B_2 \\ 0 & A & (e^{Ah} - I)B_1 & 0 \\ \hline C_1 & C_1 & 0 & D_{12} \\ C_2 & 0 & D_{21} & 0 \end{array}\right].$$

10.5 Applications

10.5.1 Parameterisation of All Stabilising Controllers

The necessary and sufficient conditions on P_h for the existence of an exponentially stabilising controller K is discussed here for the control system in Figure 10.4. It is known that exponential stabilisation for an infinite-dimensional plant is generally more difficult to achieve than stabilisation (in the H^∞ sense) [152]. It turns out that in this particular setting, the two problems are equivalent, and can be reduced to the stabilisation of $\tilde{P}_0$ from (10.14).

Theorem 10.1. *The dead-time plant P_h shown in Figure 10.4 with a minimal realization of its rational part as in (10.10) admits an (exponentially) stabilising controller K (as defined in Section 10.4) if and only if (A, B_2) is stabilisable and (C_2, A) is detectable. With the USP Z given in (10.11), every such controller can be expressed as*

$$K = C(I - ZC)^{-1}, \qquad (10.19)$$

where C is an (exponentially) stabilising controller for $\tilde{P}_0$ defined in (10.14). Let F and L be such that $A+LC_2$ and $A+B_2F$ are stable, then every stabilising C for $\tilde{P}_0$ can be expressed as

$$C = \mathcal{F}_l(M, Q), \quad M = \left[\begin{array}{c|cc} A + B_2F + E_h^{-1}LC_2E_h & -E_h^{-1}L & B_2 \\ \hline F & 0 & I \\ -C_2E_h & I & 0 \end{array}\right] \qquad (10.20)$$

where $Q \in H^\infty$. Such a C is exponentially stabilising if and only if Q is exponentially stable. The closed-loop transfer function T_{zw} achieved by K from (10.19) is given by (10.17), where

$$\mathcal{F}_l(\tilde{P}_0, C) = \mathcal{F}_l(N, Q),$$

$$N = \left[\begin{array}{cc|cc} A+B_2F & -B_2F & E_h^{-1}B_1 & B_2 \\ 0 & A+E_h^{-1}LC_2E_h & E_h^{-1}(B_1+LD_{21}) & 0 \\ \hline C_1+D_{12}F & -D_{12}F & 0 & D_{12} \\ 0 & C_2E_h & D_{21} & 0 \end{array}\right].$$

Proof. As explained in the previous section, the control system in Figure 10.4 has an equivalent representation shown in Figure 10.6. The blocks Z_1 and $\tilde{P}_s$ in Figure 10.6 are exponentially stable, so that they have no influence on the (exponential) stability of the whole system (since they are not a part of any feedback loop). Thus, the (exponential) stability of the original closed-loop system is equivalent to that of the system formed from $\tilde{P}_0$ and C only.

At first, the stabilisation in the H^∞ sense is considered. Since $\tilde{P}_0$ is rational, the parameterisation of stabilising controllers using M follows from [181, p. 312] where C_2 is replaced by C_2E_h and L by $E_h^{-1}L$. This is possible because $A+LC_2$ is similar to $A+E_h^{-1}LC_2E_h$ (hence, $A+LC_2$ is stable if and only if $A+E_h^{-1}LC_2E_h$ is stable). For a rigorous derivation of the parameterisation for irrational plants [20]. The formula involving N follows from [181, p. 323].

Now consider the exponential stabilisation. Let $\alpha < 0$ be such that all the eigenvalues of $A+B_2F$ and $A+LC_2$ are in the half-plane where Re $s < \alpha$. Now all the arguments used to derive the parameterisation (10.20) of stabilising controllers can be redone with H^∞ replaced by $H^\infty(\mathbb{C}_\alpha)$, the space of bounded analytic functions on $\mathbb{C}_\alpha$. This will result in the same formulae (10.20), but now Q must be in $H^\infty(\mathbb{C}_\alpha)$. □

10.5.2 The H^2 Problem

Consider again the feedback system from Figure 10.4 with the plant P_h. The standard H^2 problem for P_h is to find a stabilising controller K which minimises the H^2-norm of the transfer function T_{zw}. The solution of this problem, using the modified Smith predictor, is well known; see [89, 91] and the references therein.[2] In this subsection, this problem will be re-considered using the USP. Assume that the realization of P, of the form (10.10), is minimal and assume the following:

(A1) (A, B_2) is stabilisable and (C_2, A) is detectable;
(A2) $\begin{bmatrix} A-j\omega I & B_2 \\ C_1 & D_{12} \end{bmatrix}$ has full column rank $\forall \omega \in \mathbb{R}$;
(A3) $\begin{bmatrix} A-j\omega I & B_1 \\ C_2 & D_{21} \end{bmatrix}$ has full row rank $\forall \omega \in \mathbb{R}$;
(A4) $D_{12}^*D_{12} = I$ and $D_{21}D_{21}^* = I$.

Assumption (A4) is made just to simplify the exposition. In fact, only the nonsingularity of $D_{12}^*D_{12}$ and $D_{21}D_{21}^*$ is required [40, 181].

As mentioned in Remark 10.5, the impulse response of Z_1 is supported on $[0, h]$ and the last two terms of T_{zw} in (10.17) (which together are $\mathcal{F}_l(\tilde{P}, C)$ delayed by h) are supported on $[h, \infty)$. Assume for a moment that $D_{11} = 0$. It follows that Z_1 is orthogonal (in H^2) to the last two terms of T_{zw} and that

---
[2] The case when there are multiple I/O delays can be found in [95].

$$\|T_{zw}\|_2^2 = \|Z_1\|_2^2 + \left\|\mathcal{F}_l(\tilde{P}, C)e^{-sh}\right\|_2^2$$
$$= \|Z_1\|_2^2 + \left\|\mathcal{F}_l(\tilde{P}, C)\right\|_2^2.$$

Note that Z_1 is independent of C. Therefore, the standard H^2 problem for P_h, minimising $\|T_{zw}\|_2$ over all stabilising K, is reduced to

$$\gamma = \min_C \left\|\mathcal{F}_l(\tilde{P}, C)\right\|_2 \quad (C \text{ stabilising}). \tag{10.21}$$

This is a finite-dimensional standard H^2 problem whose solution is well known. The problem (10.21) is meaningful even if $D_{11} \neq 0$: then to minimise the L^2 norm of the impulse response of T_{zw} restricted to $[h, \infty)$.

It is worth discussing the solution of (10.21) because the solution is simpler than one would expect from the general solution presented in [40, Chapter 5] or [181, Chapter 14]. Rewrite $\tilde{P}$ from (10.18) as

$$\tilde{P} = \left[\begin{array}{c|cc} \tilde{A} & \tilde{B}_1 & \tilde{B}_2 \\ \hline \tilde{C}_1 & 0 & D_{12} \\ \tilde{C}_2 & D_{21} & 0 \end{array}\right].$$

The classical solution of (10.21), the standard H^2 problem for the system $\tilde{P}$, involves the Riccati equations corresponding to the Hamiltonian matrices

$$\tilde{H} = \begin{bmatrix} \tilde{A} & 0 \\ -\tilde{C}_1^*\tilde{C}_1 & -\tilde{A}^* \end{bmatrix} - \begin{bmatrix} \tilde{B}_2 \\ -\tilde{C}_1^*D_{12} \end{bmatrix} \begin{bmatrix} D_{12}^*\tilde{C}_1 & \tilde{B}_2^* \end{bmatrix},$$

$$\tilde{J} = \begin{bmatrix} \tilde{A}^* & 0 \\ -\tilde{B}_1\tilde{B}_1^* & -\tilde{A} \end{bmatrix} - \begin{bmatrix} \tilde{C}_2^* \\ -\tilde{B}_1 D_{21}^* \end{bmatrix} \begin{bmatrix} D_{21}\tilde{B}_1^* & \tilde{C}_2 \end{bmatrix}.$$

The Riccati equation for $\tilde{H}$, in the unknown $\tilde{X}$, is

$$\begin{bmatrix} -\tilde{X} & I \end{bmatrix} \tilde{H} \begin{bmatrix} I \\ \tilde{X} \end{bmatrix} = 0,$$

and similarly for $\tilde{J}$; see the books cited earlier.

However, it turns out that for the special structure of $\tilde{P}$, the standard H^2 problem can be reduced to the solution of two Riccati equations with unknowns of lower dimensions, as the following theorem shows. These reduced order Riccati equations correspond to the following two Hamiltonian matrices H_2 and J_2, which are also of lower dimensions than $\tilde{H}$ and $\tilde{J}$:

$$H_2 = \begin{bmatrix} A & 0 \\ -C_1^*C_1 & -A^* \end{bmatrix} - \begin{bmatrix} B_2 \\ -C_1^*D_{12} \end{bmatrix} \begin{bmatrix} D_{12}^*C_1 & B_2^* \end{bmatrix},$$

$$J_2 = \begin{bmatrix} A^* & 0 \\ -B_1 B_1^* & -A \end{bmatrix} - \begin{bmatrix} C_2^* \\ -B_1 D_{21}^* \end{bmatrix} \begin{bmatrix} D_{21}B_1^* & C_2 \end{bmatrix}.$$

When the conditions (A1)–(A4) hold, these two Hamiltonian matrices belong to $dom(Ric)$. Indeed, this follows from Corollary 13.10 in [181, p. 340]. The fact that $H_2 \in dom(Ric)$ means that (A, B_2) is stabilisable and H_2 has no imaginary eigenvalue. The stabilising solution of the Riccati equation associated with H_2, $X = Ric(H_2)$, satisfies $X \geq 0$. Similar statements can be made about J_2 and $Y = Ric(J_2)$.

Theorem 10.2. *If the conditions (A1)–(A4) hold, then there exists a unique optimal H^2 controller K for the plant P_h, given by*

$$K = C(I - ZC)^{-1}, \quad C = \left[\begin{array}{c|c} \tilde{A} + \tilde{B}_2\tilde{F} + \tilde{L}\tilde{C}_2 & -\tilde{L} \\ \hline \tilde{F} & 0 \end{array}\right], \qquad (10.22)$$

where

$$\tilde{F} = \left[F \ FV \begin{bmatrix} 0 \\ I_s \end{bmatrix}\right], \quad \tilde{L} = \left[\begin{bmatrix} 0 & e^{A_s h} \end{bmatrix} \begin{bmatrix} E_h^{-1}L \\ -I_s \end{bmatrix} V^{-1}L\right],$$

with

$$F = -(B_2^* X + D_{12}^* C_1), \quad L = -(YC_2^* + B_1 D_{21}^*).$$

Proof. It can be verified that the conditions (A1)–(A4) hold also for $\tilde{P}$. This follows from the zeros in the matrix in (10.18), using the fact that A_s is stable. The unique optimal H^2 controller can be obtained, according to Theorem 14.7 from [181, p. 385], as given in (10.22) but with

$$\tilde{F} = -(\tilde{B}_2^* \tilde{X} + D_{12}^* \tilde{C}_1), \quad \tilde{L} = -(\tilde{Y}\tilde{C}_2^* + \tilde{B}_1 D_{21}^*),$$

where $\tilde{X} = Ric(\tilde{H})$ and $\tilde{Y} = Ric(\tilde{J})$.

Now establish the relationship between F and $\tilde{F}$. It is known from [25, 181] that $\tilde{F}$ is the optimal state feedback gain for the standard LQ problem for

$$P_F = \left[\begin{array}{cc|c} A & 0 & B_2 \\ 0 & A_s & 0 \\ \hline C_1 & C_1 V \begin{bmatrix} 0 \\ I_s \end{bmatrix} & D_{12} \end{array}\right].$$

After a similarity transformation with

$$T_F = \begin{bmatrix} I & -V \begin{bmatrix} 0 \\ I_s \end{bmatrix} \\ 0 & I_s \end{bmatrix},$$

i.e., replacing the state x_F of P_F by $z_F = T_F^{-1} x_F$, the system P_F becomes

$$P_F' = \left[\begin{array}{cc|c} A & 0 & B_2 \\ 0 & A_s & 0 \\ \hline C_1 & 0 & D_{12} \end{array}\right],$$

because $T_F \tilde{A} = \tilde{A} T_F$. The invariant subspace corresponding to the block A_s is unobservable (and uncontrollable). Hence, the stabilising solution to the ARE for the corresponding LQ problem has the form $\begin{bmatrix} X & 0 \\ 0 & 0_s \end{bmatrix}$, where the subscript s indicates that the dimensions of 0_s are the same as those of A_s, and where $X = Ric(H_2)$. This gives the corresponding optimal state feedback gain as

$$-\begin{bmatrix} B_2^* & 0 \end{bmatrix} \begin{bmatrix} X & 0 \\ 0 & 0_s \end{bmatrix} - D_{12}^* \begin{bmatrix} C_1 & 0 \end{bmatrix} = -\begin{bmatrix} B_2^* X + D_{12}^* C_1 & 0 \end{bmatrix} = \begin{bmatrix} F & 0 \end{bmatrix}.$$

Going back to the original state of P_F via the similarity transformation with T_F^{-1}, the optimal feedback gain for the LQ problem for P_F is

$$\tilde{F} = \begin{bmatrix} F & 0 \end{bmatrix} T_F^{-1} = \begin{bmatrix} F & FV \begin{bmatrix} 0 \\ I_s \end{bmatrix} \end{bmatrix}.$$

Now establish the relationship between L and $\tilde{L}$. Note that $\tilde{L}^*$ is the optimal state feedback gain for the standard LQ problem for the plant

$$P_L = \left[\begin{array}{cc|c} A^* & 0 & E_h^* C_2^* \\ 0 & A_s^* & 0 \\ \hline B_1^* E_h^{-*} & B_1^* V^{-*} \begin{bmatrix} 0 \\ e^{A_s^* h} - I_s \end{bmatrix} & D_{21}^* \end{array} \right].$$

Using the similarity transformation with

$$T_L = \begin{bmatrix} E_h^* & -E_h^* V^{-*} \begin{bmatrix} 0 \\ e^{A_s^* h} - I_s \end{bmatrix} \\ 0 & I_s \end{bmatrix},$$

i.e., replacing the state x_L of P_L by $z_L = T_L^{-1} x_L$, the system P_L becomes

$$P_L' = \left[\begin{array}{cc|c} A^* & 0 & C_2^* \\ 0 & A_s^* & 0 \\ \hline B_1^* & 0 & D_{21}^* \end{array} \right],$$

because $T_L \tilde{A}^* = \tilde{A}^* T_L$. The invariant subspace corresponding to the block A_s^* is unobservable (and uncontrollable). Hence, the stabilising solution to the ARE for the corresponding LQ problem has the form $\begin{bmatrix} Y & 0 \\ 0 & 0_s \end{bmatrix}$, where $Y = Ric(J_2)$. This gives the corresponding optimal state feedback gain as

$$-\begin{bmatrix} C_2 & 0 \end{bmatrix} \begin{bmatrix} Y & 0 \\ 0 & 0_s \end{bmatrix} - D_{21} \begin{bmatrix} B_1^* & 0 \end{bmatrix} = \begin{bmatrix} L^* & 0 \end{bmatrix}.$$

Going back to the original state of P_L via the similarity transformation with T_L^{-1}, the optimal state feedback gain for the LQ problem for P_L is

$$\tilde{L}^* = \begin{bmatrix} L^* & 0 \end{bmatrix} T_L^{-1} = \begin{bmatrix} L^* E_h^{-*} & L^* V^{-*} \begin{bmatrix} 0 \\ e^{A_s^* h} - I_s \end{bmatrix} \end{bmatrix}.$$

Hence,

$$\tilde{L} = T_L^{-*} \begin{bmatrix} L \\ 0 \end{bmatrix} = \begin{bmatrix} E_h^{-1} L \\ \begin{bmatrix} 0 & e^{A_s h} - I_s \end{bmatrix} V^{-1} L \end{bmatrix}.$$

□

10.5.3 A Transformed H^∞ Problem

According to (10.17), the following formula holds:

$$\|Z_1\|_\infty < \|T_{zw}\|_\infty \leq \|Z_1\|_\infty + \left\|\mathcal{F}_l(\widetilde{P},C)\right\|_\infty. \tag{10.23}$$

Hence, the minimisation of $\left\|\mathcal{F}_l(\widetilde{P},C)\right\|_\infty$ may offer a very good approximation for the minimisation of $\|T_{zw}\|_\infty$, in particular, when $\|Z_1\|_\infty$ is large. This indicates that the original H^∞ problem min $\|T_{zw}\|_\infty$ may be approximately solved by solving a transformed H^∞ problem [161]:

$$\min_C \left\|\mathcal{F}_l(\widetilde{P},C)\right\|_\infty. \tag{10.24}$$

As a matter of fact, this is reasonable: during the period $t = 0 \sim h$ after w is applied, the output z is *not* controllable (*i.e.*, not changeable by the control action) and is *only* determined by Z_1 (and, of course, w). It does not make sense to include an uncontrollable part in a performance index [161] and hence it should be excluded. See Chapter 8 for more details about this idea and [68] for an application to the damping control of inter-area oscillations of large-scale power systems.

The solution to the transformed problem (10.24) involves the following Hamiltonian matrices:

$$H_\infty = \begin{bmatrix} \tilde{A} & \gamma^{-2}\tilde{B}_1\tilde{B}_1^* \\ -\tilde{C}_1^*\tilde{C}_1 & -\tilde{A}^* \end{bmatrix} - \begin{bmatrix} \tilde{B}_2 \\ -\tilde{C}_1^*D_{12} \end{bmatrix} \begin{bmatrix} D_{12}^*\tilde{C}_1 & \tilde{B}_2^* \end{bmatrix},$$

$$J_\infty = \begin{bmatrix} \tilde{A}^* & \gamma^{-2}\tilde{C}_1^*\tilde{C}_1 \\ -\tilde{B}_1\tilde{B}_1^* & -\tilde{A} \end{bmatrix} - \begin{bmatrix} \tilde{C}_2^* \\ -\tilde{B}_1 D_{21}^* \end{bmatrix} \begin{bmatrix} D_{21}\tilde{B}_1^* & \tilde{C}_2 \end{bmatrix}.$$

Assume that the assumptions (A1)–(A4) in the previous subsection hold, then the following result holds:

Theorem 10.3. *There exists a controller C such that $\left\|\mathcal{F}_l(\widetilde{P},C)\right\|_\infty < \gamma$ iff the following three conditions hold:*
  (i) $H_\infty \in dom(Ric)$ and $X = Ric(H_\infty) \geq 0$;
  (ii) $J_\infty \in dom(Ric)$ and $Y = Ric(J_\infty) \geq 0$;
  (iii) $\rho(XY) < \gamma^2$.
Moreover, when the above conditions hold, the set of all admissible controllers C such that $\left\|\mathcal{F}_l(\widetilde{P},C)\right\|_\infty < \gamma$ can be parameterised as

$$C = \mathcal{F}_l(M,Q), \tag{10.25}$$

with $Q(s) \in H^\infty$, $\|Q(s)\|_\infty < \gamma$ and

$$M = \left[\begin{array}{c|cc} A_\infty & -L_\infty \tilde{B}_2 + \gamma^{-2} Y \tilde{C}_1^* D_{12} \\ F_\infty \Psi & 0 & I \\ \hline -\left(\tilde{C}_2 + \gamma^{-2} D_{21} \tilde{B}_1^* X\right)\Psi & I & 0 \end{array}\right],$$

where

$$A_\infty = \tilde{A} + L_\infty \tilde{C}_2 + \gamma^{-2} Y \tilde{C}_1^* \tilde{C}_1 + \left(\tilde{B}_2 + \gamma^{-2} Y \tilde{C}_1^* D_{12}\right) F_\infty \Psi,$$

$$F_\infty = -(\tilde{B}_2^* X + D_{12}^* \tilde{C}_1), \quad L_\infty = -(Y\tilde{C}_2^* + \tilde{B}_1 D_{21}^*), \quad \Psi = (I - \gamma^{-2} Y X)^{-1}.$$

Furthermore, the H^∞ norm of T_{zw} satisfies (10.23) when the controller K is designed as $K = C(I - ZC)^{-1}$ with Z given in (10.11) and C given in (10.25).

Proof. As shown in the proof of Theorem 10.2, $\tilde{P}(s)$ meets all the standard assumptions. Theorem 5.1 in [39] can be used to solve the problem. Substitute $\tilde{P}(s)$ into the theorem, then the above result can be obtained with ease. □

Remark 10.7. In this case, the predictor Z is much simpler than that resulting from the original H^∞ problem; see (7.11). The price paid for this can be estimated from (10.23). The original H^∞ problem incorporating a USP, min $\|T_{zw}\|_\infty$, is left for future research.

10.6 Summary

A numerical problem with the modified Smith predictor when the plant has fast stable poles has been pointed out and the unified Smith predictor has been proposed as a solution. An equivalent representation of the augmented plant consisting of a dead-time plant and a unified Smith predictor is derived. Using this representation, a parameterisation of the (exponentially) stabilising controllers for the dead-time plant (with the USP connected to it) is derived and the standard H^2 problem (again, in the presence of the USP) and a transformed H^∞ control problem are solved. The H^∞ control problems in the presence of the USP are left for future research.

Part II

Controller Implementation

11

Discrete-delay Implementation of Distributed Delay in Control Laws

As shown in the previous chapters, suboptimal controllers for the Nehari problem, the extended Nehari problem, the one-block problem and the standard problem have the same structure. They all incorporate a distributed-delay block, which is in the form of a modified Smith predictor (MSP). The implementation of distributed delay is not trivial because of the inherent hidden unstable poles. In this chapter,[1] some elementary mathematical tools are used to approximate the distributed delay and to implement it in the z-domain and in the s-domain. The H^∞-norm of the approximation error converges to 0 when the number N of approximation steps approaches $+\infty$. Hence, the instability problem due to the approximation error (which has been widely studied in recent years) does not exist provided that the number N of approximation steps is large enough. Moreover, the static gain is guaranteed in the implementation so that no extra effort is needed to retain the steady-state performance. It is recommended not to use the backward rectangular rule to approximate the distributed delay for implementation. As by-products, two new formulae for the forward and backward rectangular rules are obtained. These formulae are more accurate than the conventional ones when the integrand has an exponential term.

11.1 Introduction

Distributed delays (*i.e.*, finite integrals over time, also called finite-impulse-response FIR blocks) often appear as part of dead-time compensators for processes with dead time, in particular, for unstable processes with dead time [70, 113, 147]. They also appear in H^∞ control of (even, stable) dead-time systems [80, 87, 160, 161, 162] and continuous-time deadbeat control [159]. Due to the requirement of internal stability, such an FIR block has to be, approximately, implemented as a stable block without hidden unstable poles.

---

[1] Portions reprinted, with permission, from [167]. ©IEEE.

One way to approximate the distributed delay is by an easy-to-implement rational function. See the next two chapters for more details about this.

A common way is to replace the distributed delay by the sum of a series of discrete (often commensurate) delays [70, 113, 147] (other interesting implementations using resetting mechanisms can be found in [135] and [96]). However, it has emerged recently that this approximation method (more specifically, using quadrature rules such as rectangular, trapezoidal and Simpson's rules etc.) cannot guarantee system stability, even when quite accurate approximation integral laws were used [141]. This topic has received much attention from the delay community and has become a very hot topic in recent years [27, 28, 81, 86, 96, 98, 123, 141, 142]. It has been proposed as an open problem in the survey paper [121]. An analysis of the causes of such behaviour was studied in [27, 123, 141] using a simple example. It was shown in [27] that the resulting system becomes a neutral time-delay system and closed-loop poles having large magnitude located in the right-half plane (whatever the precision of the trapezoidal approximation) caused the instability.

It is now well understood that the existence of a low-pass filter in the approximation may remedy the instability problem, as explicitly or implicitly reported in [28, 81, 86]. Indeed, this is a standard technique to convert a neutral time-delay system into a retarded time-delay system; see, for example, [153]. However, it is not clear why the approximation, which involves only the classical quadrature rules and Laplace transform, has lost the inherent low-pass property of the distributed delay and caused instability, nor is it clear how to choose a suitable low-pass filter (see the last paragraph of [86, Subsection 4.2]). This chapter intends to answer these questions and proposes some improved approximations to implement the distributed delay (an *approximation* is called an *implementation* only when it is implementable).

In the literature, this problem is often considered in the context of a control system. It often involves change of the control structure, *e.g.*, due to an algebraic loop, inserting a low-pass filter or even the redesign of a control law. Here, this problem is regarded as a pure approximation/implementation problem in the frequency domain. Two different reasonings will be applied, but the results obtained are the same. In the proposed implementation, both the low frequency behaviour and the high frequency behaviour are guaranteed. Moreover, the H^∞-norm of the approximation error converges to 0 when the approximation step N approaches $+\infty$. Hence, there is no change of control structure; there is no instability problem provided that N is large enough. As shown in simulations, a widely studied system [27, 123, 141], which demonstrated instability, is stable even when $N = 1$.

A bad approximation of the distributed delay discussed in the literature is reviewed in Section 11.2. Two different approaches are then proposed in Section 11.3. Based on the approximations obtained, the implementations in the z-domain and in the s-domain are given in Section 11.4. The stability issue is discussed in Section 11.5 and numerical examples are given in Section 11.6.

11.2 A Bad Approximation of Distributed Delay in the Literature

Finite-spectrum assignment [70, 109, 144] for dead-time systems is a state feedback control law using the predicted state of the control plant. See Chapter 2. The predicted state of the plant $P(s)e^{-sh}$ with $P = \left[\begin{array}{c|c}A & B \\ \hline C & 0\end{array}\right]$ is

$$x_p(t) = e^{Ah}x(t) + \int_0^h e^{A\zeta}Bu(t-\zeta)d\zeta, \qquad (11.1)$$

and the FSA control law is given by

$$u(t) = \tilde{F}e^{-Ah}x_p(t).$$

Denote the integral in (11.1), which is a distributed delay, by

$$v(t) = \int_0^h e^{A\zeta}Bu(t-\zeta)d\zeta. \qquad (11.2)$$

Using the Laplace transform, the s-domain equivalent $Z(s)$ of the distributed delay (11.2), i.e., the transfer function from u to v, is

$$Z(s) = (I - e^{-(sI-A)h}) \cdot \left[\begin{array}{c|c}A & B \\ \hline I & 0\end{array}\right]. \qquad (11.3)$$

Hence, in the frequency domain, the *distributed* delay Z is a system including a *discrete* delay but with a special property that all poles are cancelled by its zeros, i.e., it is an entire function. This paves the way for the techniques mentioned in [114] to be applied to approximate Z with a rational function, with special attention paid to avoiding the unstable poles.

The integral (distributed delay) $v(t)$ from (11.2) can be approximated in the time domain by using various quadrature rules such as rectangular, trapezoidal and Simpson's rules *etc.* In this chapter, the analysis is based on the rectangular rule for simplicity. The approximated $v(t)$ using the backward rectangular rule is

$$v_w(t) = \frac{h}{N}\sum_{i=1}^{N} e^{iA\frac{h}{N}} Bu(t - i\frac{h}{N}), \qquad (11.4)$$

where N is the number of approximation steps. The Laplace transformation of $v_w(t)$ gives the following approximation of $Z(s)$ (i.e., the transfer function from u to v_w):

$$Z_w(s) = \frac{h}{N} \cdot \sum_{i=1}^{N} e^{-i\frac{h}{N}(sI-A)} B. \qquad (11.5)$$

Z_w is not a good implementation of Z [86, 123, 141]. A simple reason is that the original FIR Z is strictly proper but Z_w is not. The bad approximation at high frequencies makes the stability analysis *unnecessarily* complicated and, what is worse, does not guarantee the stability [27, 81, 98, 123, 141].

11.3 Approximation of Distributed Delay

11.3.1 Integration $\int_0^{\frac{h}{N}} y(t-\tau)d\tau$

Lemma 11.1. *For any integrable function $y(t)$ and the rectangular pulse function*

$$p(t) = 1(t) - 1(t - \frac{h}{N}),$$

where $1(t)$ is the step function, the following identity holds:

$$\int_0^{\frac{h}{N}} y(t-\tau)d\tau = \int_{t-\frac{h}{N}}^{t} y(\tau)d\tau = y(t) * p(t),$$

where $*$ stands for the convolution.

Proof. The first "=" is obvious. The second "=" can be proved using the definition of the convolution. This formula is illustrated in Figure 11.1. □

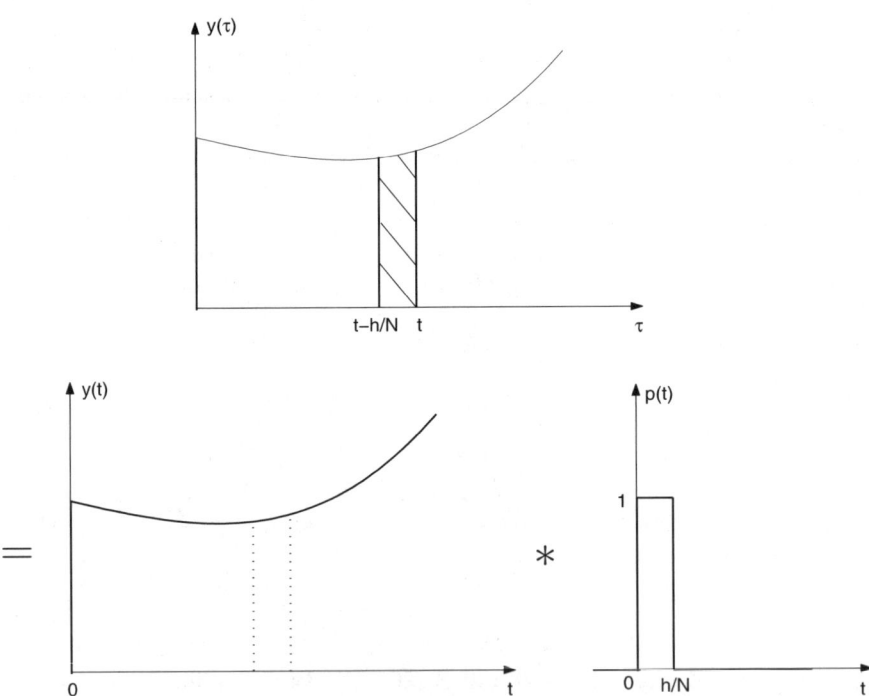

Figure 11.1. Illustration of Lemma 11.1

This formula seems trivial, but it reveals the secret behind the problem. Obviously, the Laplace transformation of $\int_0^{\frac{h}{N}} y(t-\tau)d\tau$ is $Y(s) \cdot \frac{1-e^{-s\frac{h}{N}}}{s}$, where $Y(s)$ is the Laplace transformation of $y(t)$, and $\frac{1-e^{-s\frac{h}{N}}}{s}$ is that of $p(t)$. However, when a quadrature rule is used to approximate $\int_0^{\frac{h}{N}} y(t-\tau)d\tau$, then the corresponding Laplace transformation is $Y(s)$ multiplied by a polynomial of delays. For example, when the forward rectangular rule is used, i.e.,

$$\int_0^{\frac{h}{N}} y(t-\tau)d\tau \approx y(t) \cdot \frac{h}{N},$$

this polynomial is $\frac{h}{N}$; when the backward rectangular rule is used, this polynomial is $\frac{h}{N}e^{-\frac{h}{N}s}$; when the trapezoidal rule is used, this polynomial is $\frac{h}{N}\frac{1+e^{-\frac{h}{N}s}}{2}$. Hence, in the frequency domain, the quadrature approximation can be interpreted as approximating $\frac{1-e^{-s\frac{h}{N}}}{s}$ with a polynomial of delay $e^{-\frac{h}{N}s}$. This approximation loses the strict properness of $\frac{1-e^{-s\frac{h}{N}}}{s}$. An approximation, which does not lose this important property, will be given in Subsection 11.4.2.

11.3.2 Approximation in the s-domain via the Laplace Transform

Divide the interval $[0, h]$ into N subintervals $[i\frac{h}{N}, (i+1)\frac{h}{N}]$, $i = 0, 1, \ldots, N-1$, then $v(t)$ given in (11.2) can be re-written as

$$v(t) = \sum_{i=0}^{N-1} \int_{i\frac{h}{N}}^{(i+1)\frac{h}{N}} e^{A\zeta} Bu(t-\zeta)d\zeta. \tag{11.6}$$

When N is chosen large enough, $e^{A\zeta}$ in the interval $[i\frac{h}{N}, (i+1)\frac{h}{N}]$ can be well approximated by $e^{iA\frac{h}{N}}$. This offers the following approximation for $v(t)$:

$$v(t) \approx v_f(t) = \sum_{i=0}^{N-1} e^{iA\frac{h}{N}} B \int_{i\frac{h}{N}}^{(i+1)\frac{h}{N}} u(t-\zeta)d\zeta$$

$$= \sum_{i=0}^{N-1} e^{iA\frac{h}{N}} B \int_0^{\frac{h}{N}} u(t-i\frac{h}{N}-\tau_i)d\tau_i, \tag{11.7}$$

where the variable changes $\zeta = \tau_i + i\frac{h}{N}$ were used and the subscript "f" stands for forward. This approximation can also be obtained by applying the technique used in [28], which involves block-pulse functions. The reasoning used here is simpler and needs less mathematical background.

Applying Lemma 11.1, the last formula becomes

$$v_f(t) = \sum_{i=0}^{N-1} e^{iA\frac{h}{N}} Bu(t-i\frac{h}{N}) * p(t). \tag{11.8}$$

On the other hand, if it is assumed that

$$u(t - i\frac{h}{N} - \tau_i) = u(t - i\frac{h}{N}) \quad \text{for} \quad 0 \le \tau_i < \frac{h}{N}, \tag{11.9}$$

then v_f given in (11.7) can be further approximated as

$$v_f(t) \approx v_{wf}(t) = \frac{h}{N} \sum_{i=0}^{N-1} e^{iA\frac{h}{N}} Bu(t - i\frac{h}{N}). \tag{11.10}$$

This is exactly the approximation of v by using the forward rectangular rule. As will be shown later, the approximation v_f does not cause instability when N is large enough. However, as is known, the approximation v_{wf} does. The significant difference between v_f and v_{wf} is the convolution with $p(t)$. An alternative interpretation is that the condition (11.9) is not explicitly shown in (11.10), which was used to apply the Laplace transform. The Laplace transformation of (11.8) gives the following approximation of $Z(s)$ (i.e., the transfer function from u to v_f):

$$Z_f(s) = \frac{1 - e^{-s\frac{h}{N}}}{s} \cdot \sum_{i=0}^{N-1} e^{-i\frac{h}{N}(sI-A)} B. \tag{11.11}$$

Similarly, the transfer function from u to v_{wf} is

$$Z_{wf}(s) = \frac{h}{N} \sum_{i=0}^{N-1} e^{-i\frac{h}{N}(sI-A)} B. \tag{11.12}$$

Theorem 11.1. *The approximation Z_f has the following properties:*
(i) $\lim_{N \to +\infty} Z_f(s) = Z(s)$;
(ii) Z_f is strictly proper, i.e., $\lim_{|s| \to +\infty, \Re(s) \ge 0} Z_f(s) = 0$.

Proof.

$$\begin{aligned}
\lim_{N \to +\infty} Z_f(s) &= \lim_{N \to +\infty} \frac{1 - e^{-s\frac{h}{N}}}{s} \cdot \sum_{i=0}^{N-1} e^{-i\frac{h}{N}(sI-A)} B \\
&= \lim_{N \to +\infty} \frac{1 - e^{-s\frac{h}{N}}}{s} (I - e^{-(sI-A)h})(I - e^{-\frac{h}{N}(sI-A)})^{-1} B \\
&= (I - e^{-(sI-A)h}) \lim_{N \to +\infty} \frac{1 - e^{-s\frac{h}{N}}}{s} (I - e^{-\frac{h}{N}(sI-A)})^{-1} B \\
&= (I - e^{-(sI-A)h}) \lim_{\tau \to 0^+} \frac{1 - e^{-s\tau}}{s} (I - e^{-\tau(sI-A)})^{-1} B \\
&= (I - e^{-(sI-A)h}) \lim_{\tau \to 0^+} e^{-s\tau} \cdot ((sI - A)e^{-\tau(sI-A)})^{-1} B \\
&= (I - e^{-(sI-A)h})(sI - A)^{-1} B \\
&= Z(s),
\end{aligned}$$

where the substitution $\tau = h/N$ is used. The second property is obvious. □

11.3 Approximation of Distributed Delay

Remark 11.1. Actually, $\lim_{N\to+\infty} Z_{wf}(s) = Z(s)$ as well. However, Z_{wf} is not strictly proper. This makes the stability analysis of the system, as done in the literature, very complicated. Furthermore, as will be proved later, Z_f converges to Z uniformly.

Remark 11.2. Roughly speaking, if the convolution of $p(t)$ is regarded as a hold effect to reconstruct the signal obtained from the rectangular rule, then the hold effect in (11.10) can be interpreted as the holder being a constant gain h/N, for which the impulse response is a weighted delta function $\frac{h}{N} \cdot \delta(t)$. See Figure 11.3(a) for more details.

Although the approximation (11.11) guarantees the high-frequency behaviour of the distributed delay Z, the approximation error at low frequencies might be large. In particular, the nonzero error at the zero frequency is not desirable. It changes the system performance at the steady state, as can be seen from the simulations in [141] (also see Figure 11.8), and hence extra effort to guarantee the steady-state performance is needed [147]. This means a change of the control law, as used in [86, Example 2], is needed. Such efforts can be eliminated by using a different approximation as follows.

Instead of approximating $e^{A\zeta}$ with $e^{iA\frac{h}{N}}$ as in (11.7), it can be approximated with the mean value of $e^{A\zeta}$ in the interval. This offers the following approximation:[2]

$$v(t) \approx v_{f0}(t) = \sum_{i=0}^{N-1} \frac{N}{h} \int_{i\frac{h}{N}}^{(i+1)\frac{h}{N}} e^{A\zeta} d\zeta \cdot B \cdot \int_{i\frac{h}{N}}^{(i+1)\frac{h}{N}} u(t-\zeta)d\zeta$$

$$= (e^{\frac{h}{N}A} - I)(\frac{h}{N}A)^{-1} \cdot v_f(t).$$

The corresponding approximation of Z in the s-domain is given by

$$Z_{f0}(s) = \frac{1-e^{-\frac{h}{N}s}}{s} \frac{e^{\frac{h}{N}A} - I}{\frac{h}{N}} A^{-1} \cdot \sum_{i=0}^{N-1} e^{-i\frac{h}{N}(sI-A)} B. \qquad (11.13)$$

Theorem 11.2. Z_{f0} *has the following properties:*
(i) $\lim_{N\to+\infty} Z_{f0}(s) = Z(s)$;
(ii) Z_{f0} *is strictly proper;*
(iii) $\lim_{s\to 0} Z_{f0}(s) = \lim_{s\to 0} Z(s)$.

---

[2] If A is singular, then an appropriate limitation should be used to calculate some elements of $\frac{N}{h}(e^{\frac{h}{N}A} - I)A^{-1}$ when necessary. As a matter of fact,

$$\frac{N}{h}(e^{\frac{h}{N}A} - I)A^{-1} = I + \frac{1}{2!}\frac{h}{N}A + \frac{1}{3!}(\frac{h}{N}A)^2 + \cdots + \frac{1}{(n+1)!}(\frac{h}{N}A)^n + \cdots$$

and hence it is well defined for a singular A as well. It can also be replaced by the integral $\frac{N}{h}\int_0^{\frac{h}{N}} e^{A\zeta}d\zeta$, which is nonsingular for a singular A. Similar situations are for $(I-e^{Ah})A^{-1}$ and $\frac{N}{h}(I-e^{-\frac{h}{N}A})A^{-1}$.

Proof. Property (i) is obvious since $\lim_{N\to+\infty} \frac{N}{h}(e^{\frac{h}{N}A} - I)A^{-1} = I$. Property (ii) is obvious as well.

The static gain of Z_{f0} is the same as that of Z because

$$\lim_{s\to 0} Z_{f0}(s) = \lim_{s\to 0} \frac{1 - e^{-s\frac{h}{N}}}{s} \frac{e^{\frac{h}{N}A} - I}{\frac{h}{N}} A^{-1} \sum_{i=0}^{N-1} e^{-i\frac{h}{N}(sI-A)} B$$

$$= (e^{\frac{h}{N}A} - I)A^{-1} \sum_{i=0}^{N-1} e^{i\frac{h}{N}A} B$$

$$= -(I - e^{Ah})A^{-1} B$$

$$= \lim_{s\to 0} Z(s).$$

□

The approximation Z_{f0} guarantees a small approximation error at both low and high frequencies, in particular, zero error at the frequencies 0 and $+\infty$. Hence, Z_{f0} is more accurate than Z_f, in particular, at low frequencies. This indicates that a similar change in the rectangular rule, which is given below, may provide better accuracy for numerical computations of integrals:

$$\int_0^h e^{A\zeta} Bu(t - \zeta) d\zeta \approx (e^{\frac{h}{N}A} - I)A^{-1} \cdot \sum_{i=0}^{N-1} e^{iA\frac{h}{N}} Bu(t - i\frac{h}{N}). \quad (11.14)$$

When $A = 0$, $(e^{\frac{h}{N}A} - I)A^{-1}$ becomes $\frac{h}{N}$ and hence this new formula can be regarded as an extension of the conventional forward rectangular rule. It provides a better approximation than the conventional forward rectangular rule when the integrand has an exponential term.

The magnitude coefficient $\frac{N}{h}(e^{\frac{h}{N}A} - I)A^{-1}$ depends on N (more accurately, on Ah/N). In the scalar case, it approaches 1 when $Ah/N \to 0$ (this can be extended to the matrix case). The magnitude coefficient curve is shown in Figure 11.2 for a scalar A. The larger the Ah/N, the greater the improvement in accuracy. Another magnitude coefficient $\frac{N}{h}(I - e^{-\frac{h}{N}A})A^{-1}$, which will be encountered later, is also shown in Figure 11.2.

11.3.3 Direct Approximation in the s-domain

Since the $Z(s)$ in (11.3) can be written as

$$Z(s) = \int_0^h e^{-(sI-A)\theta} d\theta \cdot B,$$

the new formula (11.14) provides the true value for Z as[3]

---

[3] Another way is to use the formula $(1 - a)\sum_{i=0}^{N-1} a^i = (1 - a^N)$.

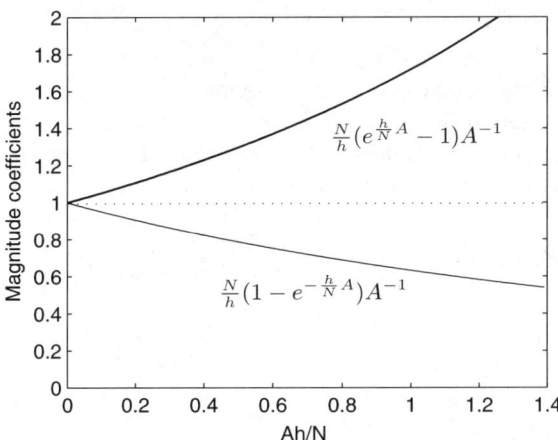

Figure 11.2. Magnitude coefficients (for a scalar A)

$$Z(s) = (I - e^{-(sI-A)\frac{h}{N}})(sI - A)^{-1} \cdot \sum_{i=0}^{N-1} e^{-i\frac{h}{N}(sI-A)} B. \qquad (11.15)$$

Hence, the distributed delay Z has been converted to the sum of a series of discrete delays, although there still exist hidden unstable poles in the first part denoted as

$$H(A) = (I - e^{-(sI-A)\frac{h}{N}})(sI - A)^{-1}.$$

It is relatively easy to implement $H(A)$ because it involves a much shorter delay h/N when N is large.

The above $H(A)$ is a function of matrix A. It can be expanded as the following power series of A:

$$H(A) = \frac{1 - e^{-\frac{h}{N}s}}{s} I + \frac{1 - e^{-\frac{h}{N}s} - \frac{h}{N}se^{-\frac{h}{N}s}}{s^2} \cdot A$$
$$+ \frac{1}{2!} \frac{2(1 - e^{-\frac{h}{N}s}) - \frac{h}{N}s(2 + \frac{h}{N}s)e^{-\frac{h}{N}s}}{s^3} \cdot A^2 + \cdots.$$

This power series (uniformly) converges for any square matrix A, provided that $H(sI)$ is defined to be $\frac{h}{N}I$ [61]. The approximation of the first term

$$H(A) \approx \frac{1 - e^{-\frac{h}{N}s}}{s} I$$

provides the following approximation of $Z(s)$:

$$Z_f(s) = \frac{1 - e^{-s\frac{h}{N}}}{s} \cdot \sum_{i=0}^{N-1} e^{-i\frac{h}{N}(sI-A)} B.$$

This is the same as the Z_f obtained in (11.11). Note that the hold filter $\frac{1-e^{-\frac{h}{N}s}}{s}$ appears again, although the reasoning used here is different. As a matter of fact, Z can be regarded as a generalised holder with a period of h (because the impulse response of Z is nonzero only in $[0, h]$).

Furthermore, the coefficients in the series of $H(A)$ can be separated into the sum of a term not involving s and a term including factors of s, i.e.,

$$H(A) = \frac{1-e^{-\frac{h}{N}s}}{s} \cdot \left(I + \frac{1}{2!}\frac{h}{N}A + \frac{1}{3!}(\frac{h}{N}A)^2 + \cdots\right.$$

$$+ (\frac{1 - \frac{s}{1-e^{-\frac{h}{N}s}}\frac{h}{N}e^{-\frac{h}{N}s}}{s} - \frac{1}{2!}\frac{h}{N}) \cdot A$$

$$+ (\frac{1}{2!}\frac{2 - \frac{s(2+\frac{h}{N}s)}{1-e^{-\frac{h}{N}s}}\frac{h}{N}e^{-\frac{h}{N}s}}{s^2} - \frac{1}{3!}(\frac{h}{N})^2) \cdot A^2 + \cdots\right)$$

$$= \frac{1-e^{-\frac{h}{N}s}}{s} \cdot \left((e^{\frac{h}{N}A} - I)(\frac{h}{N}A)^{-1} + \hat{H}(A)\right),$$

where $\hat{H}(A)$ represents the rest of the series in the bracket above. It is easy to show that $\hat{H}(A) = 0$ when $s \to 0$ or $A = 0$. The approximation of the first term

$$H(A) \approx \frac{1-e^{-\frac{h}{N}s}}{s} \cdot (e^{\frac{h}{N}A} - I)(\frac{h}{N}A)^{-1}$$

provides the following approximation of $Z(s)$:

$$Z_{f0}(s) = \frac{1-e^{-\frac{h}{N}s}}{s} \frac{e^{\frac{h}{N}A} - I}{\frac{h}{N}} A^{-1} \cdot \sum_{i=0}^{N-1} e^{-i\frac{h}{N}(sI-A)} B.$$

This is the same as the Z_{f0} obtained in (11.13). As has been shown earlier, Z_f does not guarantee the static gain of Z, but Z_{f0} does.

11.3.4 Equivalents for the Backward Rectangular Rule

The approximations Z_f and Z_{f0} have an index range of $i = 0, \cdots, N-1$ and hence may be regarded as corresponding to the *forward* rectangular rule. Similar approximations Z_b and Z_{b0}, which have an index range of $i = 1, \cdots, N$ and correspond to the *backward* rectangular rule, are

$$Z_b(s) = \frac{1-e^{-s\frac{h}{N}}}{s} \cdot \sum_{i=1}^{N} e^{-i\frac{h}{N}(sI-A)} B, \qquad (11.16)$$

and

$$Z_{b0}(s) = \frac{1-e^{-s\frac{h}{N}}}{s} \cdot \frac{I - e^{-\frac{h}{N}A}}{\frac{h}{N}} A^{-1} \sum_{i=1}^{N} e^{-i\frac{h}{N}(sI-A)} B. \qquad (11.17)$$

Similarly to (11.14), the last formula corresponds to the following quadrature approximation formula:

$$\int_0^h e^{A\zeta} Bu(t-\zeta)d\zeta \approx (I - e^{-\frac{h}{N}A})A^{-1} \cdot \sum_{i=1}^N e^{iA\frac{h}{N}} Bu(t - i\frac{h}{N}). \quad (11.18)$$

When $A = 0$, $(I - e^{-\frac{h}{N}A})A^{-1}$ becomes $\frac{h}{N}$ and hence this new formula can be regarded as an extension of the conventional backward rectangular rule. It provides a better approximation than the conventional backward rectangular rule when the integrand has an exponential term. The curve of the magnitude coefficient $\frac{N}{h}(I - e^{-\frac{h}{N}A})A^{-1}$ is shown in Figure 11.2 for a scalar A. The larger the Ah/N, the more the improvement of the accuracy.

The following results hold (with proofs omitted) for Z_b and Z_{b0}:

Theorem 11.3. *The approximation Z_b has the following properties:*
(i) $\lim_{N \to +\infty} Z_b(s) = Z(s)$;
(ii) Z_b is strictly proper.

Theorem 11.4. *The approximation Z_{b0} has the following properties:*
(i) $\lim_{N \to +\infty} Z_{b0}(s) = Z(s)$;
(ii) Z_{b0} is strictly proper;
(iii) $\lim_{s \to 0} Z_{b0}(s) = \lim_{s \to 0} Z(s)$.

There exists a pure one-step delay $e^{-\frac{h}{N}s}$ in Z_b and Z_{b0} (because i starts from 1 to N). It turns out that dropping this term improves the approximation. As a matter of fact, when the pure delay term $e^{-\frac{h}{N}s}$ in Z_{b0} is dropped, Z_{b0} becomes the same as Z_{f0} (see the simulations in Section 11.6 for the accuracy comparison). Hence, the backward rectangular rule is not recommended to implement a distributed delay. The implementation of Z in the next section and the stability issue in Section 11.5 will be done for Z_f and Z_{f0} only, although some simulations will be given in Section 11.6 for comparison.

11.4 Implementation of Distributed Delay Z

11.4.1 Implementation of Z in the z-domain

The approximations Z_f and Z_{f0}, given in (11.11) and (11.13), incorporate a hold filter $\frac{1 - e^{-\frac{h}{N}s}}{s}$. This is nothing other than a zero-order hold (ZOH), which is an element normally existing in a sampled-data system, and the rest is approximately a polynomial of z^{-1}, by using $z \approx e^{s\frac{h}{N}}$. Hence, these transfer functions can be approximately implemented in the z-domain with a sampling period of $\frac{h}{N}$, as shown[4] in Figure 11.3. As pointed out by Kannai

---
[4] Only the implementation Z_{f0} is needed because Z_f does not guarantee the static gain. It is given here for comparison with Z_{wf} in (11.12).

and Weiss ([53], Proposition 4.1), the implementations shown in Figure 11.3 converge to the corresponding transfer functions when $N \to +\infty$. Hence, this approximation step does not change the system stability, provided that N is large enough.

It is worth noting that the ZOH block implemented in SIMULINK® has two special properties: (i) the static gain is unity and thus a gain $\frac{h}{N}$ in the series of the ZOH block is needed to implement $\frac{1-e^{-\frac{h}{N}s}}{s}$; (ii) it incorporates a sampler as well and thus it is not a pure ZOH.[5] The implementation of Z_f (when ignoring the S and ZOH blocks) looks very similar to Z_{wf} in (11.12), but there is a significant difference: Z_{wf} in (11.12) is in the s-domain but the implementation in Figure 11.3 is in the z-domain. When Z is implemented in the z-domain, the resulting system is a hybrid system and the stability cannot be analysed by simply replacing the delay term z^{-1} with $e^{-\frac{h}{N}s}$. What has been done here is actually the digital implementation of a continuous time control law. Another way is to re-design a controller for the sampled plant, as reported in [142].

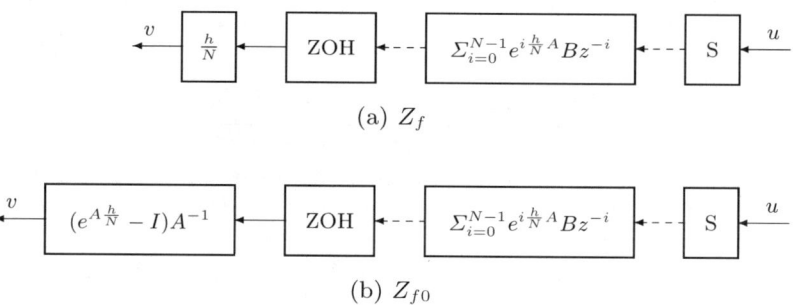

Figure 11.3. Implementations of Z in the z-domain

11.4.2 Implementation of ZOH in the s-domain

Z_{f0} does not include any hidden unstable poles of the plant, i.e., the eigenvalues of A. However, if Z_{f0} is to be implemented in the s-domain, extra care has to be taken over the implementation of the hold filter because it includes a hidden unstable pole at $s = 0$. This is more or less the same as the original problem but it is much easier to remove the hidden unstable pole $s = 0$ because the delay $\frac{h}{N}$ involved can be made much shorter than the original delay

---

[5] This will change the system behaviour when using Z_f or Z_{f0} for $N = 1$ because, in this case, the polynomial of z^{-1} is 1 and only a pure ZOH cascaded with a constant gain is needed. See the next subsection for more details.

h (which implies the approximation is to be made in a much shorter period, in the sense of the impulse response) and the pole $s = 0$ is known.

The hold filter can be expanded as the following series in ϵ:

$$\frac{1-e^{-\frac{h}{N}s}}{s} = \frac{1-e^{-\frac{h}{N}(s+\epsilon)}}{s+\epsilon} + \frac{1-e^{-\frac{h}{N}(s+\epsilon)} - \frac{h}{N}(s+\epsilon)e^{-\frac{h}{N}(s+\epsilon)}}{(s+\epsilon)^2}\epsilon + \cdots,$$

and hence it can be approximated by the first term as

$$\frac{1-e^{-\frac{h}{N}s}}{s} \approx \frac{1-e^{-\frac{h}{N}(s+\epsilon)}}{s+\epsilon}.$$

Here, $\epsilon > 0$ is chosen to be a small number close to 0 and hence $\frac{1}{s+\epsilon}$ is stable and implementable. Similarly to before, this approximation does not guarantee the static gain, but the following one does:

$$\frac{1-e^{-\frac{h}{N}s}}{s} \approx \frac{1-e^{-\frac{h}{N}(s+\epsilon)}}{s+\epsilon} \frac{\frac{h}{N}\epsilon}{1-e^{-\epsilon h/N}}. \tag{11.19}$$

This implementation of ZOH, which will be used to implement Z in the next subsection, is in the form of approximating an unstable FIR block with a stable FIR block as described in [147].

The rest of this subsection makes it clear that the ZOH implemented in MATLAB®/SIMULINK® has unity static gain and includes a sampler. Figure 11.4(a) shows the impulse response of the ZOH block implemented in MATLAB®/SIMULINK® and Figure 11.4(b) shows the implementation (11.19), with $\epsilon = 0.01$ and the hold period $\frac{h}{N} = 1$. The width of the input pulse is 4×10^{-6} s and the amplitude is 2.5×10^5. Due to the sampler included in the ZOH, the impulse response in Figure 11.4(a) has an amplitude of 2.5×10^5, which is the sampled value of the pulse amplitude. However, the impulse response of the implementation (11.19) has an amplitude of 1, as expected. Moreover, it does not depend on the amplitude of the input pulse. The width of the impulse responses for both implementations are the same as the hold period (it is 1 s in the figures). Apparently, the ZOH block implemented in MATLAB®/SIMULINK® should not be used when a pure ZOH is needed.

11.4.3 Implementation of Z in the s-domain

Using the implementation of the hold filter as in (11.19), the distributed delay Z can now be *implemented* in the s-domain, corresponding to Z_{f0}, as

$$Z_{f\epsilon}(s) = \frac{1-e^{-\frac{h}{N}(s+\epsilon)}}{1-e^{-\frac{h}{N}\epsilon}} \frac{e^{\frac{h}{N}A} - I}{s/\epsilon + 1} A^{-1} \cdot \Sigma_{i=0}^{N-1} e^{-i\frac{h}{N}(sI-A)} B. \tag{11.20}$$

Theorem 11.5. *The implementation $Z_{f\epsilon}$ has the following properties:*
  (i) $\lim_{N \to +\infty} Z_{f\epsilon}(s) = Z(s)$;
  (ii) $Z_{f\epsilon}$ *is strictly proper;*
  (iii) $\lim_{s \to 0} Z_{f\epsilon}(s) = \lim_{s \to 0} Z(s)$.

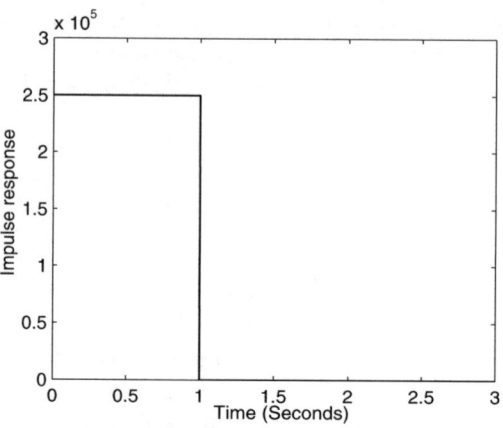

(a) implementation in MATLAB®/SIMULINK®

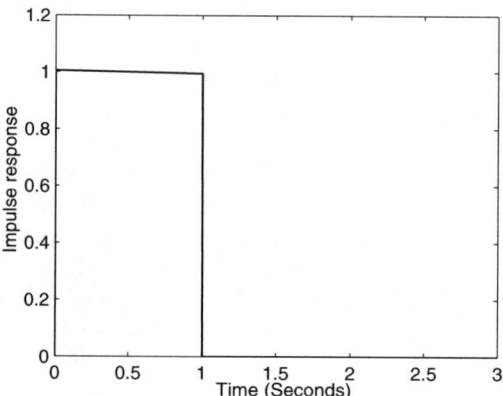

(b) proposed implementation (11.19)

Figure 11.4. Impulse responses of different implementations of ZOH

Proof.

$$\lim_{N\to+\infty} \frac{1-e^{-\frac{h}{N}(s+\epsilon)}}{1-e^{-\frac{h}{N}\epsilon}} \frac{1}{s/\epsilon+1} = \lim_{\tau\to 0^+} \frac{1-e^{-\tau(s+\epsilon)}}{1-e^{-\tau\epsilon}} \frac{1}{s/\epsilon+1}$$
$$= \lim_{\tau\to 0^+} \frac{(s+\epsilon)e^{-\tau(s+\epsilon)}}{\epsilon e^{-\tau\epsilon} \cdot (s/\epsilon+1)}$$
$$= 1,$$

where the substitution $\tau = h/N$ was used. As a result, $\lim_{N\to+\infty} Z_{f\epsilon}(s) = \lim_{N\to+\infty} Z_{wf}(s) = Z(s)$. The second property is obvious and the last one is easy to prove. □

Remark 11.3. Here, ϵ is a small positive number. It can be chosen as close to 0 as possible whenever it is implementable. However, there is no simple guideline to choose the low-pass filter for the strictly proper implementation in [86]; see the last paragraph of [86, Subsection 4.2]. For implementation by adding a low-pass filter proposed in [81], no suggestions were made for the choice of low-pass filter.

Remark 11.4. The low-pass filter in the implementations proposed in [81, 86] is added artificially to remedy the instability. The low-pass filter in (11.20) is inherently there.

11.5 Stability Issues Related to the Implementation

Denote the approximation error of Z_f as

$$E_f = Z_f - Z,$$

and similarly for the other approximation errors. As explained earlier, the approximation errors of Z_f, Z_{f0} and $Z_{f\epsilon}$ can be made as small as desirable by choosing a large enough number N. Crucially, they are all *strictly proper*. This makes the well-known small-gain theorem [40, 181] applicable for stability analysis. Otherwise, a more complicated notion, w-stability [34, 35], is needed.

Theorem 11.6. *The following formulae hold:*

$$\lim_{N \to +\infty} \|E_f(s)\|_\infty = 0,$$

$$\lim_{N \to +\infty} \|E_{f\epsilon}(s)\|_\infty = 0, \qquad (\epsilon \geq 0).$$

Proof. Only the first formula will be proved. The second one can be proved similarly. According to (11.11) and (11.15), E_f is given by

$$E_f(s) = \left(\frac{1 - e^{-s\frac{h}{N}}}{s} I - (I - e^{-(sI-A)\frac{h}{N}})(sI - A)^{-1} \right) \cdot \sum_{i=0}^{N-1} e^{-i\frac{h}{N}(sI-A)} B$$

$$= E_1(s) Z_{wf}(s),$$

where Z_{wf} is as given in (11.12) and

$$E_1(s) = \frac{N}{h} \left(\frac{1 - e^{-s\frac{h}{N}}}{s} I - (I - e^{-(sI-A)\frac{h}{N}})(sI - A)^{-1} \right)$$

$$= \frac{N}{h} \int_0^{\frac{h}{N}} e^{-s\tau} (I - e^{A\tau}) d\tau. \qquad (11.21)$$

Since $\|Z_{wf}\|_\infty$ is bounded on the closed right half-plane, it is sufficient to show that $\|E_1\|_\infty$ approaches 0 when $N \to +\infty$. $E_1(s)$ is stable and hence it is only needed to consider the convergence on the $j\omega$-axis. It is easy to see from (11.21) that

$$\|E_1(s)\|_\infty \leq \frac{N}{h} \int_0^{\frac{h}{N}} \|I - e^{A\tau}\| \, d\tau.$$

The right side approaches 0 when $N \to +\infty$. □

Remark 11.5. However, neither Z_w nor Z_{wf} has this property. This is the reason why the closed-loop system studied in the literature is unstable.

According to the small-gain theorem, the approximation/implementation error does not cause any instability when N is large enough and there is *no need for any further complicated analysis for system stability*. Such a need lies in looking for the minimal N to guarantee the system stability. This is a topic left for future research.

11.6 Numerical Examples

Consider the simple plant

$$\dot{x}(t) = x(t) + u(t-1)$$

with the control law

$$u(t) = -(1 + \lambda_d) \left(e^1 \cdot x(t) + \int_0^1 e^\zeta u(t-\zeta) d\zeta \right) + r(t). \tag{11.22}$$

This example has been widely studied in the literature [27, 28, 123, 141]. Here, $A = 1$, $B = 1$, $h = 1$ and $F = -(1 + \lambda_d)$. The closed-loop system has only one pole at $s = -\lambda_d$. The closed-loop system is stable when $\lambda_d > 0$.

11.6.1 Approximations and Implementations of Distributed Delay

In this example, the distributed delay is

$$v(t) = \int_0^1 e^\zeta u(t-\zeta) d\zeta. \tag{11.23}$$

The ideal implementation Z in the s-domain, as given by (11.3), is $Z(s) = \frac{1-e^{1-s}}{s-1}$. The implementation Z_w studied in the literature, as given by (11.5), is $Z_w(s) = \frac{1}{N} \cdot \Sigma_{i=1}^N e^{-\frac{i}{N}(s-1)}$. The following approximations (not implementations) have been discussed:

$$Z_b(s) = \frac{1 - e^{-\frac{1}{N}s}}{s} \cdot \Sigma_{i=1}^N e^{-\frac{i}{N}(s-1)},$$

$$Z_{b0}(s) = \frac{1 - e^{-\frac{1}{N}s}}{s} (1 - e^{-\frac{1}{N}}) N \cdot \Sigma_{i=1}^N e^{-\frac{i}{N}(s-1)},$$

$$Z_f(s) = \frac{1 - e^{-\frac{1}{N}s}}{s} \cdot \Sigma_{i=0}^{N-1} e^{-\frac{i}{N}(s-1)},$$

$$Z_{f0}(s) = \frac{1 - e^{-\frac{1}{N}s}}{s}(e^{\frac{1}{N}} - 1)N \cdot \Sigma_{i=0}^{N-1} e^{-\frac{i}{N}(s-1)}.$$

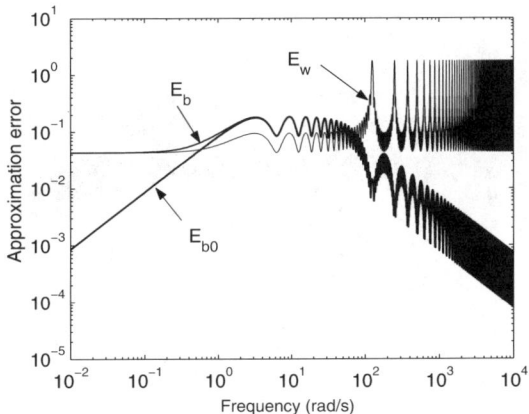

(a) E_b, E_{b0} and E_w

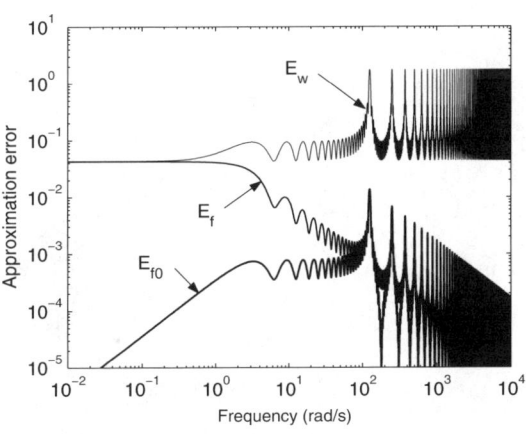

(b) E_f, E_{f0} and E_w

Figure 11.5. Errors of different approximations ($N = 20$)

The magnitude frequency responses of the approximation errors for the above approximations are shown in Figure 11.5 for $N = 20$. The approximation Z_w is not strictly proper and hence the error has a very large magnitude (it does not vanish even when $N \to +\infty$) when the frequency approaches

$+\infty$. The proposed approximations are strictly proper and the approximation errors are small. Z_{b0} and Z_{f0} guarantee the static gain of Z (*i.e.*, the error is 0 at zero frequency) so that the steady-state performance of the system [147] is guaranteed. Z_f and Z_{f0} are much better than Z_b and Z_{b0} for the same N.

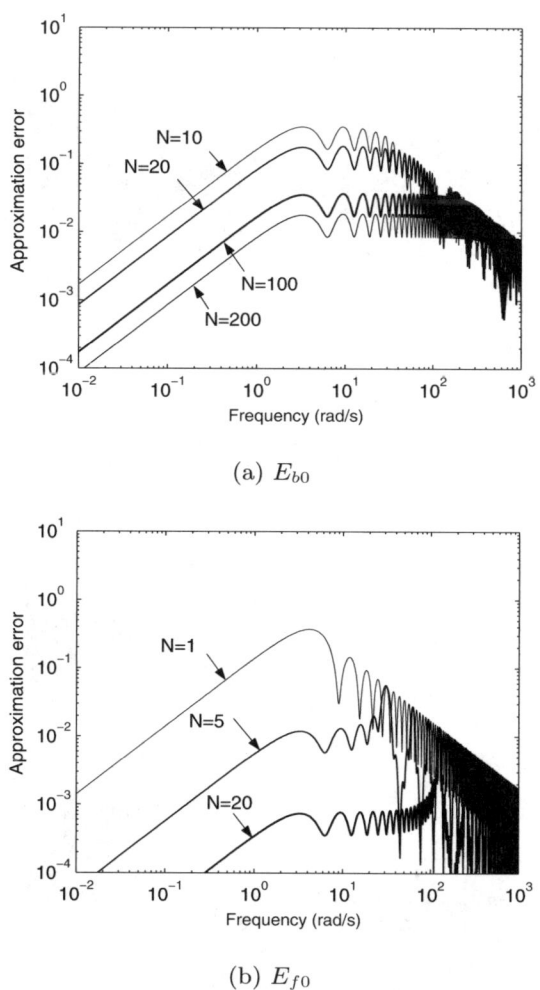

Figure 11.6. E_{b0} and E_{f0} for different N (zero error at frequencies 0 and $+\infty$)

The approximation errors of Z_{b0} and Z_{f0} for different N are shown in Figure 11.6. The larger the value of N, the smaller the approximation error. This verifies that there always exists a number N such that the stability of the closed-loop system is guaranteed. Moreover, for a certain approximation error bound, the number N required by Z_{f0} is much smaller than that required by Z_{b0} (as mentioned earlier, a simple explanation for this is that the backward rectangular rule involves an extra one-step delay h/N but the forward one does not). Z_{f0} also converges faster than Z_{b0}.

Figure 11.7 shows the implementation error of $Z_{f\epsilon}$ when $N = 1$ for $\epsilon = 1$, $0.5, 0.1$ and 0. The smaller the ϵ, the better the implementation. When $\epsilon = 0.1$, the implementation error is very close to that when $\epsilon = 0$ and there is no need to use an ϵ smaller than 0.1.

The recommended s-domain implementation of Z is the $Z_{f\epsilon}$ given in (11.20) and the z-domain implementation is the Z_{f0} shown in Figure 11.3(b).

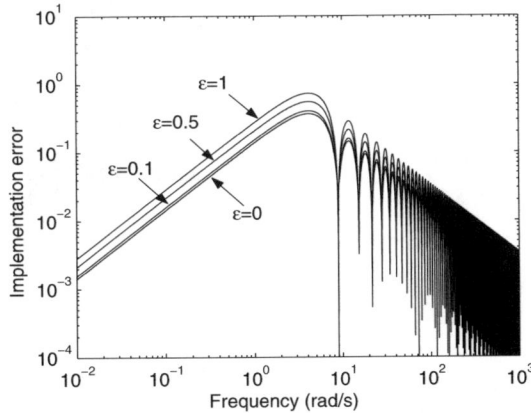

Figure 11.7. Implementation error of $Z_{f\epsilon}$ for different ϵ ($N = 1$)

11.6.2 System Responses using Different Implementations

This subsection shows the responses when $r(t) = 1$ for different implementations of Z.

In order to make comparisons with results shown in the literature, Figure 11.8(a) shows the response when Z is implemented in the z-domain as Z_b for $N = 8$, and Figure 11.8(b) shows the response when Z is implemented as Z_w. The system is stable when Z is implemented as Z_b but is unstable when Z is implemented as Z_w (as reported in the literature). The steady-state behaviour of the system has been changed. This is because Z_b does not have the same static gain as Z.

When Z is implemented in the s-domain as $Z_{f\epsilon}$ for $N = 1$, i.e.,

$$Z_{f\epsilon}(s) = \frac{e^1 - 1}{1 - e^{-\epsilon}} \frac{1 - e^{-\epsilon}e^{-s}}{s/\epsilon + 1},$$

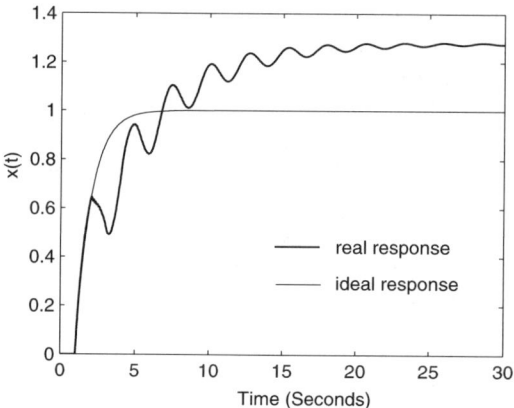

(a) Z implemented as Z_b in the z-domain

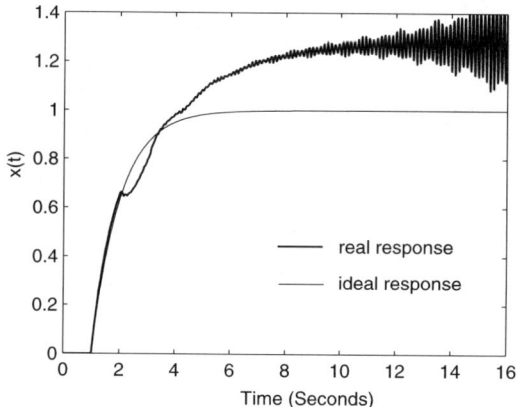

(b) Z implemented as Z_w given in (11.5)

Figure 11.8. Unit step response ($N = 8$)

Figure 11.9 shows the unit-step response of the system using the control law (11.22) with $\lambda_d = 1$, *i.e.*,

$$u = -(1 + \lambda_d)\left(e^1 \cdot x + v\right) + r, \ v = Z_{f\epsilon} \cdot u$$

in the s-domain for different ϵ (note that no change is made to the control law). No instability occurred in the simulations. The steady-state behaviour of the system is guaranteed; the transient response is slightly worse than the ideal response, which is due to the approximation of the distributed delay. For different ϵ, the smaller the ϵ (the better the approximation), the smaller the overshoot. There is no significant improvement when ϵ is less than 0.1. If a much better transient response is desired, then N has to be increased.

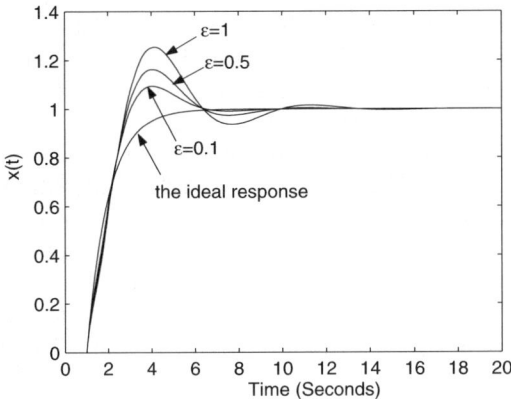

Figure 11.9. Unit-step response: Z implemented as $Z_{f\epsilon}$ $(N = 1)$

11.6.3 Numerical Integration Using the Improved Rectangular Rules

Again, consider the integral given in (11.23). Assume $N = 5$ and $u(t) = 1$ to simplify the computations. The two new formulae (11.18) and (11.14) provide the true value, $\int_0^1 e^\zeta d\zeta = e - 1$, as follows:

$$(1 - e^{-0.2}) \cdot \Sigma_{i=1}^{5} e^{0.2i} = (1 - e^{-0.2}) \cdot \frac{e - 1}{1 - e^{-0.2}} = e - 1,$$

$$(e^{0.2} - 1) \cdot \Sigma_{i=0}^{4} e^{0.2i} = (e^{0.2} - 1) \cdot \frac{1 - e}{1 - e^{0.2}} = e - 1.$$

This is because a constant $u(t)$ corresponds to zero frequency in the frequency domain, at which the approximation error is zero. However, the conventional rectangular rule provides the following approximation:

$$0.2 \cdot \Sigma_{i=1}^{5} e^{0.2i} = 0.2 \cdot \frac{e - 1}{1 - e^{-0.2}} \approx 1.10(e - 1),$$

$$0.2 \cdot \Sigma_{i=0}^{4} e^{0.2i} = 0.2 \cdot \frac{1 - e}{1 - e^{0.2}} \approx 0.90(e - 1),$$

with approximation errors of 10% and -10%, respectively.

11.7 Summary

In this chapter, it is shown that, in the frequency domain, the strict properness of the distributed delay is lost when quadrature approximations are applied. This causes the instability reported in the literature. Two different approaches are proposed to approximate the distributed delay in such a way that both low and high frequency behaviours are guaranteed and that the H^∞-norm of the approximation error converges to 0 when the number of approximation steps approaches $+\infty$. Hence, the reported instability due to the approximation error disappears and the steady-state performance of the system is also guaranteed without changing the control structure. Implementations in the z-domain and in the s-domain are proposed. As by-products, two new formulae for the forward and backward rectangular rules are obtained. These formulae are more accurate than the conventional ones when there is an exponential term in the integrand. Numerical examples have verified the results obtained.

12
Rational Implementation Inspired by the δ-operator

In this chapter,[1] a rational implementation for distributed delay is discussed. The main benefit of doing so is the easy implementation of rational transfer functions. The rational implementation, which was inspired by the δ-operator, has an elegant structure of chained low-pass filters. The stability of each node can be guaranteed by the choice of the total number N of nodes and the stability of the closed-loop system can be guaranteed because the H^∞-norm of the implementation error approaches 0 when N goes to ∞. Moreover, the steady-state performance of the system is retained without the need to change the control structure.

12.1 Introduction

In the previous chapter, a distributed delay is implemented using a series of discrete delays. Since the distributed delay is actually a block involving a discrete delay with special properties, it is possible to implement distributed delay by using rational transfer functions. The major benefit of doing so is the easy implementation of rational transfer functions. This chapter is devoted to proposing a rational implementation for distributed delay. The approach was inspired by the δ-operator, in particular, by the fact that the δ-operator approaches the differential operator when the sampling frequency approaches ∞.

The proposed implementation meets all the five key points listed below when implementing a distributed delay [171]:

1. The low-pass property of distributed delay must be kept in the implementation.
2. There is no unstable pole–zero cancellation in the implementation in order to guarantee the internal stability.
3. The implementation itself must be stable in order to guarantee the stability of the closed-loop system.

---
[1] Portions reprinted, with permission, from [171]. ©IEEE.

4. The implementation error should be able to be made small enough to guarantee the stability of the closed-loop system.
5. The static gain of the distributed delay should be retained in the implementation to guarantee the steady-state performance of the system.

The notation used in this chapter is quite standard. The left half s-plane is denoted by $\mathcal{C}_-$. A transfer function is said to be stable if all the poles are in $\mathcal{C}_-$ and a matrix A is said to be stable if all the eigenvalues are in $\mathcal{C}_-$.

12.2 The δ-operator

The δ-operator is defined as
$$\delta = \frac{q-1}{\tau},$$
where q is the shift operator and τ is the sampling period. The δ-operator creates the rapprochement between analogue and digital dynamic systems. As a matter of fact, Middleton and Goodwin unified continuous and discrete time control engineering using this operator [82, 83, 101] for more details. When $\tau \to 0$, the δ-operator approaches the differential operator, i.e., $\delta \to s$. This can be interpreted as
$$\delta = \frac{e^{\tau s} - 1}{\tau} \tag{12.1}$$
because $q \to e^{\tau s}$ when $\tau \to 0$ [53]. Re-arrange the above formula, then,
$$e^{-\tau s} = \frac{1}{\tau \delta + 1}.$$
This is not enough to approximate distributed delay due to the hidden (unstable) poles in the distributed delay. However, it is enough to inspire the approach proposed here.

12.3 An Initial Approximation

Define $\Phi = \frac{1}{\tau} I$ and, similarly to (12.1), define
$$\Delta = (e^{\tau(sI-A)} - I)\Phi, \tag{12.2}$$
then
$$sI - A = \lim_{\tau \to 0} \Delta, \tag{12.3}$$
$$e^{-(sI-A)\tau} = (\Phi^{-1}\Delta + I)^{-1}. \tag{12.4}$$
Substitute this into the s-domain equivalent (11.3) of the distributed delay (11.2), assuming $\tau = \frac{h}{N}$ with the natural number N called the number of approximation steps, then

$$Z(s) = (I - e^{-(sI-A)\tau N})(sI - A)^{-1}B$$
$$= (I - (\Phi^{-1}\Delta + I)^{-N})(sI - A)^{-1}B. \tag{12.5}$$

Using the approximation $\Delta \approx sI - A$ which can be obtained from (12.3), Z is approximated as $Z(s) \approx Z_r(s)$ with

$$Z_r(s)$$
$$= (I - (\Phi^{-1}(sI - A) + I)^{-N})(sI - A)^{-1}B \tag{12.6}$$
$$= (I - (\Phi^{-1}(sI - A) + I)^{-1}) \cdot \Sigma_{k=0}^{N-1}(\Phi^{-1}(sI - A) + I)^{-k}(sI - A)^{-1}B$$
$$= \Phi^{-1}\Sigma_{k=1}^{N}(\Phi^{-1}(sI - A) + I)^{-k}B$$
$$= \Sigma_{k=1}^{N} \Pi^k \Phi^{-1} B, \tag{12.7}$$

where
$$\Pi = (sI - A + \Phi)^{-1}\Phi.$$

The hidden, possibly unstable, poles in Z have disappeared from Z_r. This approximation converges to Z when $N \to +\infty$; see Section 12.5.

Theorem 12.1. *The approximation Z_r with $\Phi = \frac{N}{h}I$ is stable if and only if the natural number N satisfies*

$$N > \max_i \mathrm{Re}\lambda_i(A) \cdot h,$$

where $\lambda_i(A)$ is the ith eigenvalue of A.

Proof. The A-matrix of the node transfer function Π is $A - \Phi = A - \frac{N}{h}I$. In order to guarantee the stability, it is necessary and sufficient that[2] $\max_i \mathrm{Re}\lambda_i(A) - \frac{N}{h} < 0$, which gives the above condition. □

12.4 Implementation with Zero Static Error

Although there always exists a number N such that the above implementation Z_r guarantees the stability of the closed-loop system (see the next section), the static gain of $Z(s)$ is not guaranteed, i.e., $Z(0) \neq Z_r(0)$, and hence the static performance of the system cannot be retained. In order to rectify this, with a slight abuse of notation, re-define Φ as

$$\Phi = (\int_0^{\frac{h}{N}} e^{-A\zeta}d\zeta)^{-1}. \tag{12.8}$$

---

[2] The stability analysis in this chapter and the next one involves only functions of matrix A. Hence, the eigenvalues of the functions are the scalar functions of the eigenvalues of A. In other words, the eigenvalues of the functions can be obtained eigenvalue by eigenvalue. See [23, Section 4.6] for more details.

198   12 Rational Implementation Inspired by the δ-operator

The previously defined $\Phi = \frac{1}{\tau}I = \frac{N}{h}I$ can be regarded as a special case of this newly defined Φ for $A = 0$. In the sequel, unless explicitly specified, Φ is defined as in (12.8). Substitute (12.8) into (12.2), assuming $\tau = \frac{h}{N}$, then Δ defined in (12.2) becomes

$$\Delta = (e^{\tau(sI-A)} - I)\Phi = (e^{\tau(sI-A)} - I)(\int_0^\tau e^{-A\zeta}d\zeta)^{-1}. \tag{12.9}$$

This Δ satisfies the *limiting property* (12.3). Moreover, it satisfies the following *static property* as well:

$$sI - A\,|_{s=0} = \Delta\,|_{s=0} = -A.$$

Another important property of this Δ is that

$$\Delta\,|_{sI-A=0} = 0.$$

This property, called the *cancellation property*, guarantees that $(sI - A)^{-1}$ in Z can be cancelled in the approximation.

As can be easily verified, all the formulae (12.2)–(12.7) hold for this Δ as well, with Φ re-defined in (12.8). The only difference between these two cases is that the static gain can be guaranteed when Φ is defined as in (12.8), but not when $\Phi = \frac{N}{h}I$.

The stability of the node Π in this case, which is determined by the number N, is more difficult to be tested. Denote an eigenvalue of A as $\bar\sigma + j\bar\omega$. Then the corresponding eigenvalue of $\frac{h}{N}A$ is $\sigma + j\omega$ with $\sigma = \frac{h}{N}\bar\sigma$ and $\omega = \frac{h}{N}\bar\omega$.

Theorem 12.2. *The following conditions are equivalent:*
(i) Z_r, Π or $A - \Phi$ is stable;
(ii) $\int_0^{\frac{h}{N}} e^{A\zeta}d\zeta$ is antistable;
(iii) $e^{-A\frac{h}{N}}\Phi$ is antistable;
(iv) $\sigma\cos\omega + \omega\sin\omega - \sigma e^{-\sigma} > 0$, ignoring the case when $\sigma = 0$ and $\omega = 0$.

Proof. The A-matrix of the node transfer function Π is

$$A - \Phi = A - (\int_0^{\frac{h}{N}} e^{-A\zeta}d\zeta)^{-1}$$
$$= A + A(e^{-A\frac{h}{N}} - I)^{-1}$$
$$= -A(e^{A\frac{h}{N}} - I)^{-1}$$
$$= -(\int_0^{\frac{h}{N}} e^{A\zeta}d\zeta)^{-1} \tag{12.10}$$
$$= -(\int_0^{\frac{h}{N}} e^{A(\frac{h}{N}-\zeta)}d\zeta)^{-1}$$
$$= -e^{-A\frac{h}{N}}\Phi, \tag{12.11}$$

where a variable substitution $\zeta :\to \frac{h}{N} - \zeta$ was used. It was assumed that A is nonsingular in the second step. Actually, all the equalities except the second and third ones hold for singular A as well. Since the inverse operation does not change the sign of the real part of an eigenvalue, the sign of the real part of the eigenvalues of $A - \Phi$ is opposite from that of $\int_0^{\frac{h}{N}} e^{A\zeta} d\zeta$. Hence, Π is stable if and only if $\int_0^{\frac{h}{N}} e^{A\zeta} d\zeta$ is antistable, i.e.,

$$\text{Re}\lambda_i \left(\int_0^{\frac{h}{N}} e^{A\zeta} d\zeta \right) > 0. \tag{12.12}$$

According to (12.11), this is equivalent to $e^{-A\frac{h}{N}}\Phi$ being antistable. Hence, conditions (i), (ii) and (iii) are equivalent.

Apparently, the condition (12.12) holds for any eigenvalue of A with $\bar{\sigma} = 0$ and $\bar{\omega} = 0$. Hence, it is assumed that $\bar{\sigma}$ and $\bar{\omega}$ are not 0 simultaneously in the rest of the proof. The real part of the eigenvalue of $\int_0^{\frac{h}{N}} e^{A\zeta} d\zeta$ corresponding to the eigenvalue $\bar{\sigma} + j\bar{\omega}$ of A is

$$\begin{aligned}
\text{Re} \int_0^{\frac{h}{N}} e^{(\bar{\sigma}+j\bar{\omega})\zeta} d\zeta &= \int_0^{\frac{h}{N}} e^{\bar{\sigma}\zeta} \cos(\bar{\omega}\zeta) d\zeta \\
&= \frac{h}{N} \int_0^1 e^{\sigma\zeta} \cos(\omega\zeta) d\zeta \\
&= \frac{\frac{h}{N}e^{\sigma}}{\sigma^2 + \omega^2} \cdot (\sigma \cos\omega + \omega \sin\omega - \sigma e^{-\sigma}),
\end{aligned}$$

where a variable substitution $\zeta :\to \frac{h}{N}\zeta$ was used. The first term $\frac{\frac{h}{N}e^{\sigma}}{\sigma^2+\omega^2}$ is always positive. Hence, Π (or Z_r) is stable if and only if the second term is positive, i.e.,

$$\sigma \cos\omega + \omega \sin\omega - \sigma e^{-\sigma} > 0. \tag{12.13}$$

□

The surface of $f(\sigma, \omega) = \sigma \cos\omega + \omega \sin\omega - \sigma e^{-\sigma}$ is shown in Figure 12.1. The contour of $f(\sigma, \omega)$ at level 0 is shown in Figure 12.2, with the shaded area for $f(\sigma, \omega) > 0$. Since a pair of conjugate complex eigenvalues is symmetric to the real axis ($\omega = 0$), so is the contour. The surface is symmetric to the plane $\omega = 0$. When σ is very large (positive), the terms $\sigma \cos\omega + \omega \sin\omega$ dominate the surface, which oscillates; when σ is very small (negative), the term $-\sigma e^{-\sigma}$ dominates the surface, which grows up dramatically. Hence, roughly speaking, it is easier for $A - \Phi$ to be stable if A is stable. From these two figures, the choice of N to guarantee the stability of Π can be determined. In particular, put all the eigenvalues of A on Figure 12.2 and then draw lines to connect these points and the origin. N can be determined by moving these points along the lines into the shaded area. In general, the larger the delay h the larger the number N needed; the further away to the upper/lower right corner the eigenvalue of A the larger the number N needed.

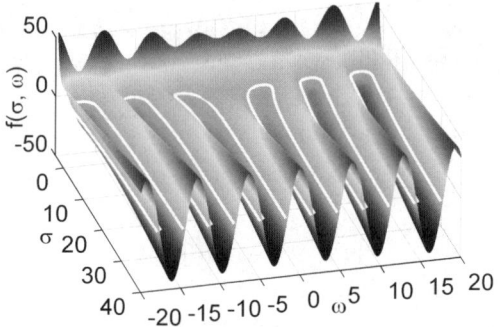

Figure 12.1. Surface of $f(\sigma, \omega)$: $f(\sigma, \omega) = 0$ on the white lines

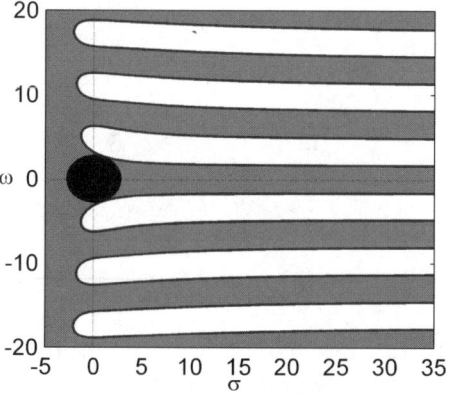

Figure 12.2. Contour of $f(\sigma, \omega)$ at level 0

Corollary 12.1. *If all the eigenvalues of A are real, then the node Π with Φ given in (12.8) is stable for any natural number N.*

Remark 12.1. Following (12.11), the node Π can be written as

$$\Pi = (sI + e^{-A\frac{h}{N}}\Phi)^{-1}\Phi = (s\Phi^{-1} + e^{-A\frac{h}{N}})^{-1}.$$

A sufficient condition is given below to facilitate the choice of N. Although this condition may be very conservative, e.g., when all eigenvalues of A are real, it offers general guidance.

Theorem 12.3. *The conditions in Theorem 12.2 are all satisfied for any number $N > \tilde{N}$ with*

$$\tilde{N} = \left\lceil \frac{h}{2.8} \cdot \max_i |\lambda_i(A)| \right\rceil, \tag{12.14}$$

where $\lceil \cdot \rceil$ is the ceiling function.

Proof. It can be proved that there is an area around the real axis, in particular a circle around the origin, such that the condition (12.13) is satisfied. See Figure 12.2. Any points on the complex plane can be scaled by $\frac{h}{N}$ into this area. This means that there always exists a number $\tilde{N}$ such that Π is stable for any N larger than $\tilde{N}$.

The minimal distance from the origin to the contour $f(\sigma, \omega) = 0$ can be numerically calculated. An approximate value is 2.8, which was measured from the circle shown in Figure 12.2. If $\max_i |\lambda_i(A)| \frac{h}{N} < 2.8$, then all eigenvalues of A can be scaled into this circle by $\frac{h}{N}$. This gives $\tilde{N}$ in (12.14). □

12.5 Convergence of the Implementation

Define
$$\Omega = \frac{N}{h}\Phi^{-1} = \frac{N}{h}\int_0^{\frac{h}{N}} e^{-A\zeta}d\zeta.$$

This matrix satisfies the following properties:
(i) $\lim_{N \to +\infty} \Omega - I = 0$,
(ii) $\lim_{N \to +\infty} N(\Omega - I) = -\frac{h}{2}A$, and
(iii) $\lim_{N \to +\infty} \Omega^2(N+1) - N \cdot I = I - Ah$.
The last two are not zero in general.

Theorem 12.4. *The following conditions are equivalent:*
(i) Ω is antistable;
(ii) Φ is antistable;
(iii) $\int_0^{\frac{h}{N}} e^{-A\zeta}d\zeta$ is antistable;
(iv) $-\sigma \cos\omega + \omega \sin\omega + \sigma e^\sigma > 0$, ignoring the case when $\sigma = 0$ and $\omega = 0$.
Furthermore, these conditions are all satisfied if $N > \tilde{N}$ with $\tilde{N}$ given in (12.14).

Proof. It is obvious that the first three conditions are equivalent. Similar to the proof of Theorem 12.2, the third condition is equivalent to the fourth one.

The contour of $-\sigma \cos\omega + \omega \sin\omega + \sigma e^\sigma$ at level 0 is symmetric to that of $f(\sigma, \omega)$ at level 0 w.r.t the imaginary axis $\sigma = 0$. Thus, the above conditions hold when (σ, ω) is in the circle shown in Figure 12.2. □

Denote the approximation error of Z_r as $E_r = Z - Z_r$. Then the following theorem holds.

Theorem 12.5. *The implementation error of Z_r satisfies*
$$\lim_{N \to +\infty} \|E_r(s)\|_\infty = 0.$$

This holds for both $\Phi = \frac{N}{h}I$ and $\Phi = (\int_0^{\frac{h}{N}} e^{-A\zeta}d\zeta)^{-1}$.

Proof. At first, consider the case with $\Phi = (\int_0^{\frac{h}{N}} e^{-A\zeta} d\zeta)^{-1}$. Substitute (11.3) and (12.6) into E_r, then

$$\begin{aligned}
E_r &= (I - e^{-(sI-A)h})(sI - A)^{-1}B - (I - (\Phi^{-1}(sI - A) + I)^{-N})(sI - A)^{-1}B \\
&= \int_0^h e^{-(sI-A)\zeta} d\zeta \cdot B - N \int_0^{\frac{h}{N}} (\Omega(sI - A)\zeta + I)^{-N-1} \Omega d\zeta \cdot B \\
&= \int_0^{\frac{h}{N}} e^{-(sI-A)N\zeta} dN\zeta \cdot B - N \int_0^{\frac{h}{N}} (\Omega(sI - A)\zeta + I)^{-N-1} \Omega d\zeta \cdot B \\
&= N \int_0^{\frac{h}{N}} \left(e^{-(sI-A)N\zeta} - (\Omega(sI - A)\zeta + I)^{-N-1} \Omega \right) B d\zeta.
\end{aligned}$$

The integrand $e^{-(sI-A)N\zeta} - (\Omega(sI - A)\zeta + I)^{-N-1} \Omega$ can be expanded into a series of ζ as

$$I - \Omega + \zeta \cdot \left(\Omega^2(N+1) - N \cdot I \right) (sI - A)(I + O(\zeta)).$$

Note that, when $N \to +\infty$, the term ζ^0 disappears but the term ζ^1 does not due to the properties of Ω given at the beginning of this section.

Now, consider the stability of the matrix

$$A - \Omega^{-1}\zeta^{-1} = A - \Phi - \Omega^{-1}(\zeta^{-1} - (\frac{h}{N})^{-1})$$

for $\zeta \in [0, \frac{h}{N}]$. When $N > \tilde{N}$, $A - \Phi$ is stable according to Theorem 12.3 and $-\Omega^{-1}$ is stable as well according to Theorem 12.4. $\zeta^{-1} - (\frac{h}{N})^{-1} > 0$ when $\zeta \in (0, \frac{h}{N}]$. Hence, $(\Omega(sI - A)\zeta + I)^{-1}$ is stable for $\zeta \in (0, \frac{h}{N}]$ when $N > \tilde{N}$. This means that the integrand is bounded on the closed right half-plane when $N > \tilde{N}$ for $\zeta \in [0, \frac{h}{N}]$ and so is $\left(\Omega^2(N+1) - N \cdot I \right) (sI - A)(I + O(\zeta))$. Assume that

$$\left\| \left(\Omega^2(N+1) - N \cdot I \right) (sI - A)(I + O(\zeta)) \cdot B \right\| < M,$$

when $N > \tilde{N}$, then

$$\|E_r(s)\|_\infty \leq \|(I - \Omega)B\| h + N \int_0^{\frac{h}{N}} M\zeta d\zeta \qquad (N > \tilde{N}). \qquad (12.15)$$

The two terms on the right-hand side all approach 0 when $N \to +\infty$. This proves the case with $\Phi = (\int_0^{\frac{h}{N}} e^{-A\zeta} d\zeta)^{-1}$.

As to the case with $\Phi = \frac{N}{h}I$, the proof is very similar. The corresponding Ω is equal to I. This considerably simplifies the proof. The lower bound of N is given in Theorem 12.1. The first term in (12.15) disappears and the second term approaches 0 when $N \to +\infty$. □

12.6 Structure of the Implementation

This theorem, together with Theorem 12.3, indicate that there always exists a number N such that the implementation is stable and, furthermore, the H^∞-norm of the implementation error is less than a given positive value. According to the well-known small-gain theorem, the stability of the closed-loop system can always be guaranteed.

12.6 Structure of the Implementation

A state-space realization of the rational implementation Z_r given in (12.7) can be found as

$$Z_r = \begin{bmatrix} A-\Phi & \Phi & 0 & \cdots & & 0 & 0 \\ 0 & A-\Phi & \ddots & \ddots & & \vdots & \vdots \\ \vdots & & \ddots & \ddots & \Phi & 0 & 0 \\ 0 & \cdots & 0 & A-\Phi & \Phi & 0 \\ 0 & \cdots & 0 & 0 & A-\Phi & I \\ \hline I & \cdots & I & I & I & 0 \end{bmatrix} B.$$

Denote

$$u_b = \Phi^{-1} B u \quad \text{and} \quad v_r = Z_r u \approx v,$$

then Z_r can be described in state-space equations as

$$\dot{x}_N = (A-\Phi)x_N + \Phi x_{N-1}$$
$$\dot{x}_{N-1} = (A-\Phi)x_{N-1} + \Phi x_{N-2}$$
$$\vdots$$
$$\dot{x}_2 = (A-\Phi)x_2 + \Phi x_1$$
$$\dot{x}_1 = (A-\Phi)x_1 + \Phi u_b$$
$$v_r = \sum_{i=1}^{N} x_i.$$

This offers an elegant structure shown in Figure 12.3. It is a chain of N low-pass filters $\Pi = (sI - A + \Phi)^{-1}\Phi$. The more nodes cascaded, the better the approximation accuracy.

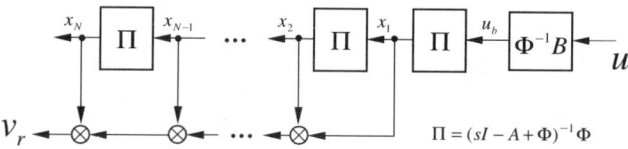

Figure 12.3. Rational implementation: $Z_r = \sum_{k=1}^{N} \Pi^k \cdot \Phi^{-1} B$

12.7 Numerical Examples

Consider the simple plant

$$\dot{x}(t) = x(t) + u(t-1)$$

with

$$u(t) = -(1+\lambda_d)\left(e^1 \cdot x(t) + \int_0^1 e^\zeta u(t-\zeta)d\zeta\right) + r(t), \qquad (12.16)$$

where $r(t)$ is the reference signal. This example has been widely studied in the literature [27, 28, 123, 141]. Here, $A = 1$, $B = 1$ and $h = 1$. The closed-loop system has only one pole at $s = -\lambda_d$, which is stable when $\lambda_d > 0$. The distributed delay in (12.16) is

$$v(t) = \int_0^1 e^\zeta u(t-\zeta)d\zeta,$$

and the s-domain equivalent is $Z(s) = (1-e^{1-s})/(s-1)$. The rational implementation Z_r with guaranteed static gain is

$$Z(s) = \Sigma_{k=1}^N \frac{\epsilon}{(\epsilon s + 1 - \epsilon)^k}, \qquad \epsilon = 1 - e^{-\frac{1}{N}}.$$

The approximation error for different N is shown in Figure 12.4. It can be seen that the static error is zero and the larger the number N the smaller the approximation error. The approximation error approaches 0 at high frequencies as well.

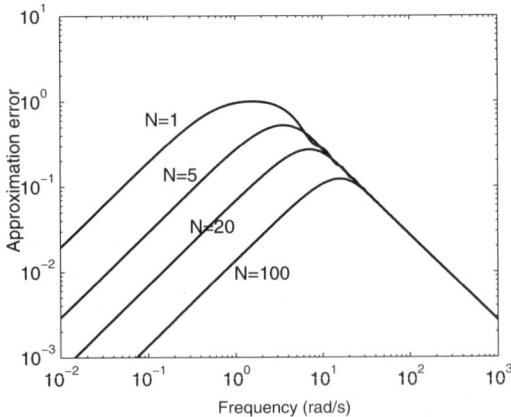

Figure 12.4. Implementation error of Z_r for different N

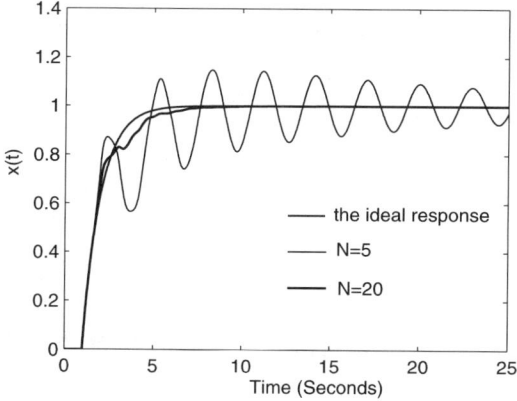

Figure 12.5. Unit-step response

Figure 12.5 shows the unit-step response of the system using the control law (12.16) with $\lambda_d = 1$, i.e.,

$$u = -(1 + \lambda_d)\left(e^1 \cdot x + v\right) + r, \; v = Z_r \cdot u$$

in the s-domain for different N (note that no change is made to the control law). When $N = 1$, the system is unstable because of the large approximation error around 1 rad/s. When $N = 5$, the system is stable but the response is oscillatory and settles down slowly. When $N = 20$, the response is very close to the ideal response. All the stable responses guarantee the steady-state performance.

12.8 Summary

Inspired from the δ-operator, an approach is proposed to implement distributed delay in linear control laws using rational transfer functions. The implementation has an elegant structure of chained low-pass filters. The criteria to guarantee the stability of each node in the chain is developed. The H^∞-norm of the implementation error approaches 0 when the number of implementation steps goes to ∞. The steady-state performance of the system is also guaranteed. Simulation examples are given to verify the results.

It turns out that the convergence is not fast enough. A fast convergent implementation will be discussed in the next chapter.

13

Rational Implementation Based on the Bilinear Transformation

Based on an extension of the bilinear transformation, a rational implementation for distributed delay in linear control laws is discussed in this chapter.[1] This implementation converges much faster than the rational implementation inspired from the δ-operator discussed in the previous chapter. The implementation has an elegant structure of chained bi-proper nodes cascaded with a strictly proper node. The stability of each node is determined by the choice of the number N of the nodes. The H^∞-norm of the implementation error approaches 0 when N goes to ∞ and hence the stability of the closed-loop system can be guaranteed. In addition, the steady-state performance of the system is retained. Simulation examples are given to verify the results and to show comparative study with other implementations.

13.1 Preliminary: Bilinear Transformation

The well-known γ-operator in digital and sampled-data control circles is

$$\gamma = \frac{2}{\tau} \cdot \frac{q-1}{q+1},$$

where q is the shift operator and τ is the sampling period [9, 33, 129]. It is often used to digitise a continuous-time transfer function. The transformation defined by the γ-operator is also called the bilinear transformation, or the Tustin's transformation. It actually corresponds to the trapezoidal rule for numerical integration. It also connects to the (lower) linear fractional transformation $\mathcal{F}_l$ and the (right) homographic transformation $\mathcal{H}_r$ with

$$\gamma = \frac{2}{\tau}\mathcal{F}_l(\begin{bmatrix} -1 & 2 \\ 1 & -1 \end{bmatrix}, q) = \frac{2}{\tau}\mathcal{H}_r(\begin{bmatrix} 1 & -1 \\ 1 & 1 \end{bmatrix}, q).$$

The term "bilinear transformation" is adopted in this chapter.

---

[1] Portions reprinted from [173], with permission from Elsevier.

The shift operator q can be solved as

$$q = \mathcal{H}_r(\begin{bmatrix} 1 & -1 \\ 1 & 1 \end{bmatrix}^{-1}, \frac{\tau}{2}\gamma) = \frac{1+\frac{\tau}{2}\gamma}{1-\frac{\tau}{2}\gamma}.$$

Since $q \to e^{\tau s}$ when $\tau \to 0$ [53], $e^{-\tau s}$ can be approximated as

$$e^{-\tau s} \approx q^{-1} = \frac{1-\frac{\tau}{2}\gamma}{1+\frac{\tau}{2}\gamma},$$

and, furthermore, γ has the following *limiting property*:

$$\lim_{\tau \to 0} \gamma = \lim_{\tau \to 0} \frac{2}{\tau} \frac{e^{\tau s} - 1}{e^{\tau s} + 1} = s.$$

This means the γ-operator is an approximation of the differential operator $p = \frac{d}{dt}$. Using the approximation $\gamma \approx s$, then $e^{-\tau s} \approx \frac{1-\frac{\tau}{2}s}{1+\frac{\tau}{2}s}$, which actually recovers the first-order Padé approximation of $e^{-\tau s}$.

Since the γ-operator offers better approximation than the δ-operator (which corresponds to the forward rectangular rule) [9, 33, 129], the γ-operator, *i.e.*, the bilinear transformation, is exploited here to implement the distributed delay.

13.2 Implementation of Distributed Delay

For a natural number N and the delay $h > 0$, the function Φ of matrix A is defined as

$$\Phi = (\int_0^{\frac{h}{N}} e^{-A\zeta} d\zeta)^{-1}(e^{-A\frac{h}{N}} + I)$$

and, furthermore, a bilinear transformation Γ is defined as

$$\Gamma = (e^{\tau(sI-A)} - I)(e^{\tau(sI-A)} + I)^{-1}\Phi, \qquad (13.1)$$

with $\tau = \frac{h}{N}$. This can be regarded as the extension of the bilinear transformation to the matrix case. Γ has the following properties:
(i) the *limiting property*

$$sI - A = \lim_{\tau \to 0} \Gamma, \qquad (13.2)$$

(ii) the *static property*

$$sI - A|_{s=0} = \Gamma|_{s=0} = -A,$$

(iii) the *cancellation property*

$$\Gamma|_{sI-A=0} = 0.$$

13.2 Implementation of Distributed Delay

According to the mechanism developed in the previous chapter, this Γ is able to bring a rational implementation for the distributed delay to guarantee the stability of the closed-loop system and the steady-state performance.

From (13.1), there is

$$e^{-(sI-A)\frac{h}{N}} = (\Phi - \Gamma)(\Phi + \Gamma)^{-1}.$$

Substitute this into (11.3), then

$$Z(s) = (I - (\Phi - \Gamma)^N (\Phi + \Gamma)^{-N})(sI - A)^{-1}B. \qquad (13.3)$$

Due to the limiting property (13.2), there is $\Gamma \approx sI - A$. Substitute this into (13.3), then Z can be approximated as Z_r given below:

$$\begin{aligned} Z_r(s) &= (I - (\Phi - sI + A)^N (sI - A + \Phi)^{-N})(sI - A)^{-1}B \qquad (13.4) \\ &= (I - (\Phi - sI + A)(sI - A + \Phi)^{-1}) \cdot \\ &\quad \Sigma_{k=0}^{N-1} (\Phi - sI + A)^k (sI - A + \Phi)^{-k} \cdot (sI - A)^{-1}B \\ &= 2(sI - A + \Phi)^{-1} \Sigma_{k=0}^{N-1} (\Phi - sI + A)^k (sI - A + \Phi)^{-k} B \\ &= \Sigma_{k=0}^{N-1} \Pi^k \Xi B, \qquad (13.5) \end{aligned}$$

with

$$\Pi = (\Phi - sI + A)(sI - A + \Phi)^{-1}, \qquad \Xi = 2(sI - A + \Phi)^{-1}.$$

The hidden, possibly unstable, poles in Z disappear from Z_r. This approximation converges to Z when $N \to +\infty$, as will be proved in Section 13.3.

Π is a bi-proper rational transfer function while Ξ is strictly proper. They share the same denominator and thus the same stability property. Z_r can be easily implemented as a chain of rational transfer functions shown in Figure 13.1. Since Ξ is strictly proper, so is Z_r. This indicates there always exists a large enough N such that the implementation does not affect the stability of the closed-loop system, provided that each node is stable. See Section 13.3.

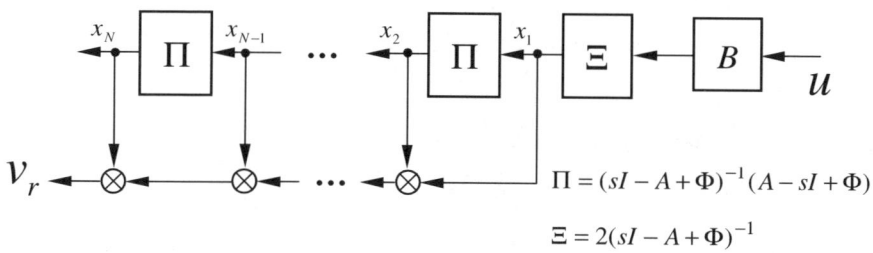

Figure 13.1. Rational implementation of distributed delay: $Z_r = \Sigma_{k=0}^{N-1} \Pi^k \Xi B$

13 Rational Implementation Based on the Bilinear Transformation

The stability of each node Π or Ξ is determined by the number N of nodes; see the theorem below. Denote an eigenvalue[2] of A as $\bar{\sigma} + j\bar{\omega}$. Then the corresponding eigenvalue of $\frac{h}{N}A$ is $\sigma + j\omega$ with $\sigma = \frac{h}{N}\bar{\sigma}$ and $\omega = \frac{h}{N}\bar{\omega}$.

Theorem 13.1. *The following conditions are equivalent:*
  (i) Z_r, Π, Ξ or $A - \Phi$ is stable;
  (ii) $\int_0^{\frac{h}{N}} e^{A\zeta} d\zeta$ is antistable;
  (iii) $\sigma \cos\omega + \omega \sin\omega - \sigma e^{-\sigma} > 0$, ignoring the case when $\sigma = 0$ and $\omega = 0$.

Proof. (i) $\Leftrightarrow$ (ii): The A-matrix of the node transfer functions is

$$A - \Phi = A - \left(\int_0^{\frac{h}{N}} e^{-A\zeta} d\zeta\right)^{-1}(e^{-A\frac{h}{N}} + I)$$
$$= A + A(e^{-A\frac{h}{N}} - I)^{-1}(e^{-A\frac{h}{N}} + I)$$
$$= 2A(e^{-A\frac{h}{N}} - I)^{-1} e^{-A\frac{h}{N}}$$
$$= -2A(e^{A\frac{h}{N}} - I)^{-1}$$
$$= -2\left(\int_0^{\frac{h}{N}} e^{A\zeta} d\zeta\right)^{-1}. \tag{13.6}$$

It was assumed that A is nonsingular here. Actually, the final equality holds for a singular A as well. Since the inverse operation does not change the sign of the real part of an eigenvalue, the signs of the real part of the eigenvalues of $A - \Phi$ are opposite from those of $\int_0^{\frac{h}{N}} e^{A\zeta} d\zeta$. Hence, Z_r (Π and Ξ) is stable if and only if $\int_0^{\frac{h}{N}} e^{A\zeta} d\zeta$ is antistable, i.e.,

$$\mathrm{Re}\lambda_i\left(\int_0^{\frac{h}{N}} e^{A\zeta} d\zeta\right) > 0. \tag{13.7}$$

(ii) $\Leftrightarrow$ (iii) has been proved in the previous chapter. It is reiterated here for the readers' convenience.

Apparently, the condition (13.7) holds for any eigenvalue of A with $\bar{\sigma} = 0$ and $\bar{\omega} = 0$. Hence, it is assumed that $\bar{\sigma}$ and $\bar{\omega}$ are not 0 simultaneously in the rest of the proof. The real part of the eigenvalue of $\int_0^{\frac{h}{N}} e^{A\zeta} d\zeta$ corresponding to the eigenvalue $\bar{\sigma} + j\bar{\omega}$ of A is

$$\mathrm{Re}\int_0^{\frac{h}{N}} e^{(\bar{\sigma}+j\bar{\omega})\zeta} d\zeta = \int_0^{\frac{h}{N}} e^{\bar{\sigma}\zeta} \cos(\bar{\omega}\zeta) d\zeta$$
$$= \frac{h}{N}\int_0^1 e^{\sigma\zeta} \cos(\omega\zeta) d\zeta$$
$$= \frac{\frac{h}{N} e^{\sigma}}{\sigma^2 + \omega^2} \cdot (\sigma\cos\omega + \omega\sin\omega - \sigma e^{-\sigma}),$$

---

[2] The stability analysis in this chapter and the previous one involves only functions of matrix A. Hence, the eigenvalues of the functions are the scalar functions of the eigenvalues of A. In other words, the eigenvalues of the functions can be obtained one by one. See [23, Section 4.6] for more details.

where a variable substitution $\zeta :\mapsto \frac{h}{N}\zeta$ was used. The first term $\frac{\frac{h}{N}e^\sigma}{\sigma^2+\omega^2}$ is always positive. Hence, Π (or Z_r) is stable if and only if the second term is positive, i.e.,

$$\sigma \cos \omega + \omega \sin \omega - \sigma e^{-\sigma} > 0. \qquad (13.8)$$

□

Corollary 13.1. *If all the eigenvalues of A are real, then each node Π or Ξ is stable for any natural number N.*

The node or the implementation shares the same stability property with the node or the implementation derived using the δ-operator in the previous chapter. Hence, some of the results there can be applied to this implementation as well. For example, the three-dimensional surface of $f(\sigma, \omega) = \sigma \cos \omega + \omega \sin \omega - \sigma e^{-\sigma}$ can be found in Figure 12.1 and a sufficient condition to guarantee the stability of the node or the implementation can be found in Theorem 12.3. The contour of $f(\sigma, \omega)$ at level 0 was given in Figure 12.2, of which a part of the figure focusing on the circle around the origin with a radius of 2.8 is shown in Figure 13.2. All the eigenvalues of $A\frac{h}{N}$ fall into this circle when $N > \tilde{N}$ with

$$\tilde{N} = \left\lceil \frac{h}{2.8} \cdot \max_i |\lambda_i(A)| \right\rceil, \qquad (13.9)$$

where $\lceil \cdot \rceil$ is the ceiling function. When this condition holds, all the conditions in Theorem 13.1 are satisfied. In particular, each node is stable.

13.3 Convergence of the Implementation

The result below is crucial to prove the convergence of the implementation.

Lemma 13.1. *Φ is antistable if $N > \tilde{N}$ with $\tilde{N}$ given in (13.9).*

Proof. Assume, temporarily, that A does not have an eigenvalue 0. According to (13.6), there is

$$\Phi\frac{h}{N} = A\frac{h}{N}(I - e^{-A\frac{h}{N}})^{-1}(e^{-A\frac{h}{N}} + I). \qquad (13.10)$$

As mentioned before, if $N > \tilde{N}$ then all the eigenvalues of $A\frac{h}{N}$ fall into the circle around the origin with a radius of 2.8 shown in Figure 13.2. The map $\phi = c\frac{1+e^{-c}}{1-e^{-c}}$ maps all the points inside this circle into the shaded area shown in Figure 13.3, where only the mapped area for the right-half-circle is shown. The mapped area for the left-half-circle overlaps with that of the right-half-circle because the map is symmetric with respect to c. The mapped area is on the open right-half-plane. Since the eigenvalues of $\Phi\frac{h}{N}$ are mapped from the

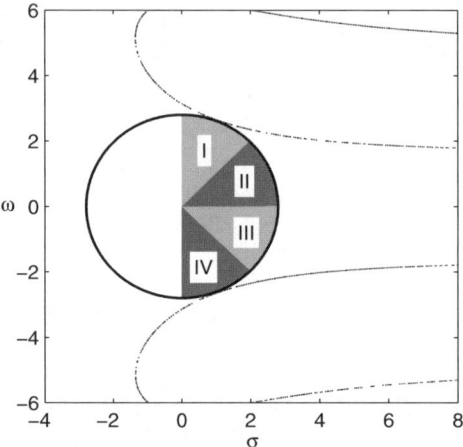

Figure 13.2. Circle into which all eigenvalues of $A\frac{h}{N}$ fall when $N > \tilde{N}$

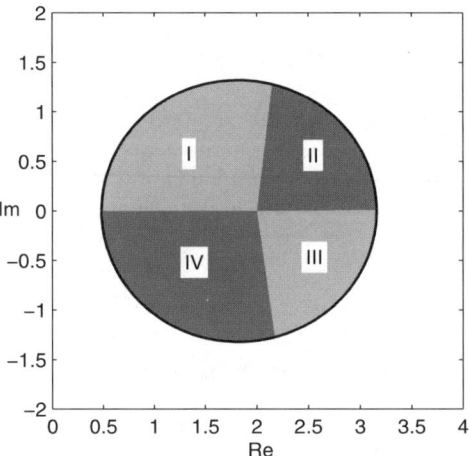

Figure 13.3. Area mapped from the right-half circle in Figure 13.2 via $\phi = c\frac{1+e^{-c}}{1-e^{-c}}$

eigenvalues of $A\frac{h}{N}$ via this map, as can be seen from (13.10), the real part of the eigenvalues of $\Phi\frac{h}{N}$ and hence of Φ is always positive. In other words, Φ is antistable when $N > \tilde{N}$.

This is also true when A has an eigenvalue of 0 (the corresponding eigenvalue of $\Phi\frac{h}{N}$ is 2) because the singularity $c = 0$ in the map ϕ is removable and the origin is mapped to the point (2, 0). □

Denote the approximation error of Z_r as $E_r = Z - Z_r$.

13.3 Convergence of the Implementation

Theorem 13.2. *The implementation error of Z_r satisfies*
$$\lim_{N \to +\infty} \|E_r(s)\|_\infty = 0.$$

Proof. According to (11.3) and (13.4), the approximation error E_r is

$$E_r = (I - e^{-(sI-A)h})(sI - A)^{-1}B -$$
$$(I - (\Phi - sI + A)^N(sI - A + \Phi)^{-N})(sI - A)^{-1}B$$
$$= \int_0^h e^{-(sI-A)\zeta} d\zeta \cdot B - 2N\Phi \cdot$$
$$\int_0^1 (-I + 2\Phi(\Phi - (sI - A)\zeta)^{-1})^{-N-1}(\Phi - (sI - A)\zeta)^{-2}d\zeta \cdot B$$
$$= \int_0^{\frac{h}{N}} e^{-(sI-A)N\zeta}dN\zeta \cdot B - N\int_0^{\frac{h}{N}} 2\Phi \cdot$$
$$(-I + 2\Phi(\Phi - (sI - A)\frac{N}{h}\zeta)^{-1})^{-N-1}(\Phi - (sI - A)\frac{N}{h}\zeta)^{-2}d\frac{N}{h}\zeta \cdot B$$
$$= N\int_0^{\frac{h}{N}} \left(e^{-(sI-A)N\zeta} - \frac{2N}{h}\Phi(-I + 2\Phi(\Phi - (sI - A)\frac{N}{h}\zeta)^{-1})^{-N-1} \cdot \right.$$
$$\left. (\Phi - (sI - A)\frac{N}{h}\zeta)^{-2} \right) d\zeta \cdot B$$
$$= N\int_0^{\frac{h}{N}} \left(e^{-(sI-A)N\zeta} - \frac{2N}{h}\Phi(\Phi + (sI - A)\frac{N}{h}\zeta)^{-N-1} \cdot \right.$$
$$\left. (\Phi - (sI - A)\frac{N}{h}\zeta)^{N-1} \right) d\zeta \cdot B.$$

The integrand can be expanded into a series of ζ as
$$I - \frac{2N}{h}\Phi^{-1} + \zeta \cdot N\left((\frac{2N}{h}\Phi^{-1})^2 - I \right)(sI - A)(I + O(\zeta)).$$

Note that, when $N \to +\infty$, both the term ζ^0 and the term ζ^1 disappear because $I - \frac{2N}{h}\Phi^{-1} \to 0$ and $N\left((\frac{2N}{h}\Phi^{-1})^2 - I\right) \to 0$. This means the convergence is much faster than the case in the previous chapter, where the term ζ^1 does not disappear when $N \to +\infty$.

Now, consider the stability of the matrix
$$A - \Phi\frac{h}{N}\zeta^{-1} = A - \Phi - \Phi((\frac{N}{h}\zeta)^{-1} - 1)$$

for $\zeta \in [0, \frac{h}{N}]$. If $N > \tilde{N}$, then $A - \Phi$ is stable, and $-\Phi$ is stable as well according to Lemma 13.1. Since $(\frac{N}{h}\zeta)^{-1} - 1 \geq 0$ when $\zeta \in (0, \frac{h}{N}]$, $(\Phi + (sI - A)\frac{N}{h}\zeta)^{-N-1}(\Phi - (sI - A)\frac{N}{h}\zeta)^{N-1}$ is stable for $\zeta \in (0, \frac{h}{N}]$ when $N > \tilde{N}$. This means that the integrand is bounded on the closed right half-plane when $N > \tilde{N}$ for $\zeta \in [0, \frac{h}{N}]$ and so is $N\left((\frac{2N}{h}\Phi^{-1})^2 - I\right)(sI - A)(I + O(\zeta))$. Assume

$$\left\| N\left((\frac{2N}{h}\Phi^{-1})^2 - I\right)(sI - A)(I + O(\varsigma))\cdot B\right\| < M, \qquad (N > \tilde{N})$$

then

$$\|E_r(s)\|_\infty \leq \left\|(I - \frac{2N}{h}\Phi^{-1})B\right\|h + N\int_0^{\frac{h}{N}} M\varsigma d\varsigma \qquad (N > \tilde{N}). \qquad (13.11)$$

The two terms on the right-hand side all approach 0 when $N \to +\infty$. □

This theorem indicates that there alway exists a number N such that the implementation is stable and, moreover, the H^∞-norm of the implementation error is less than a given positive value. According to the well-known small-gain theorem, the stability of the closed-loop system can always be guaranteed.

13.4 Numerical Examples

As in the previous two chapters, consider the simple plant $\dot{x}(t) = x(t) + u(t-1)$ with the control law

$$u(t) = -(1 + \lambda_d)\left(e^1 \cdot x(t) + \int_0^1 e^\varsigma u(t - \varsigma)d\varsigma\right) + r(t), \qquad (13.12)$$

where $r(t)$ is the reference signal. The closed-loop system has only one pole at $s = -\lambda_d$, which is stable when $\lambda_d > 0$. The distributed delay in (13.12) is

$$v(t) = \int_0^1 e^\varsigma u(t - \varsigma)d\varsigma. \qquad (13.13)$$

and the s-domain equivalent is

$$Z(s) = \frac{1 - e^{1-s}}{s - 1}.$$

The rational implementation Z_r is

$$Z_r(s) = \Sigma_{k=0}^{N-1}\left(\frac{2 - (1-\epsilon)s}{2\epsilon + (1-\epsilon)s}\right)^k \frac{2(1-\epsilon)}{2\epsilon + (1-\epsilon)s}.$$

where $\epsilon = e^{-\frac{1}{N}}$. Since A has no nonreal eigenvalues, Z_r is always stable, even for $N = 1$.

The approximation error for different N is shown in Figure 13.4. The static error is zero and the approximation error approaches 0 at both high and low frequencies. The approximation error decreases when the number N of the nodes increases. The convergence is fast, in particular, at low frequencies.

Figure 13.5 shows the implementation error of different implementations for $N = 5$. The discrete-delay implementation discussed in Chapter 11 is denoted "discrete delay" in the figure and the rational implementation discussed

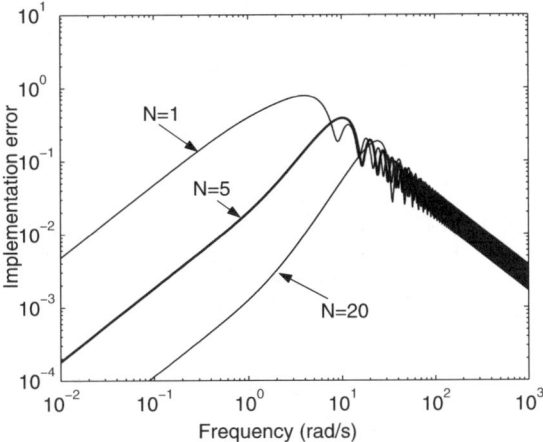

Figure 13.4. Implementation error of Z_r for different N

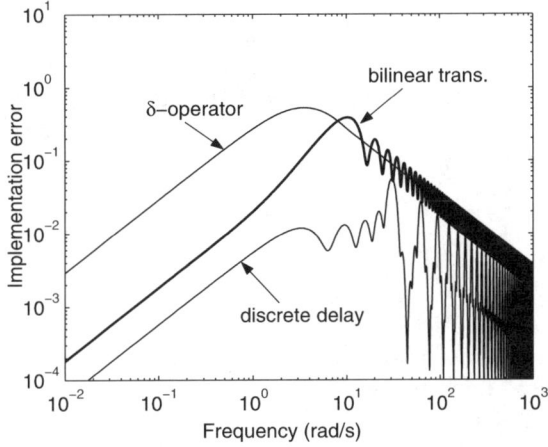

Figure 13.5. Comparison of different implementations ($N = 5$)

in Chapter 12 is denoted "δ-operator". The implementation discussed in this chapter, denoted "bilinear trans.", is much better than the one derived using the δ-operator. Although it is still worse than the discrete-delay implementation, it has the advantage of easy implementation. Actually, the proposed implementation is good enough, as can be seen from Figure 13.6, where the step response when $N = 5$ is very close to the ideal response.

The unit-step responses of the system, as shown in Figure 13.6, are obtained by using the control law (13.12) with $\lambda_d = 1$, i.e.,

$$u = -(1 + \lambda_d)\left(e^1 \cdot x + v\right) + r, \; v = Z_r \cdot u$$

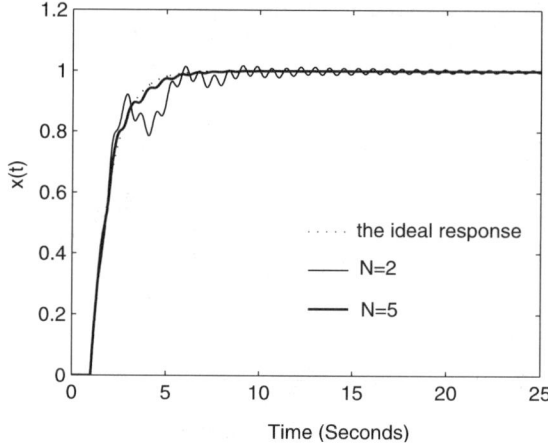

Figure 13.6. System responses when $r(t) = 1(t)$

in the s-domain for different N (note that no change is made to the control law). When $N = 1$, the system is unstable because of the large approximation error. When $N = 2$, the system is stable though slightly oscillatory. When $N = 5$, the response is very close to the ideal response. All the stable responses guarantee the steady-state performance.

13.5 Summary

Based on an extension of the bilinear transformation, an approach is introduced to implement distributed delay in linear control laws using rational transfer functions. The implementation consists of a series of bi-proper nodes cascaded with a low-pass node. The implementation converges much faster than the one proposed in the previous chapter. It has been proved that the H^∞-norm of the implementation error approaches 0 when the number N of nodes goes to ∞. Hence, there always exists a number N to guarantee the stability of the closed-loop system. In addition, the steady-state performance of the system is also guaranteed. In addition to the easy implementation, it does not involve any additional parameter to choose apart from the number N of the nodes. In particular, no parameter for a low-pass filter is needed to choose, which is an essential part in the literature [81, 86, 97, 167]. Simulation examples are given to verify the results and to compare different implementations.

References

[1] A. Amstutz and L.R. Del Re. EGO sensor based robust output control of EGR in diesel engines. *IEEE Trans. Control Syst. Technology*, 3(1):39–48, 1995.

[2] K.H. Ang, G. Chong, and Y. Li. PID control system analysis, design, and technology. *IEEE Trans. Control Syst. Technology*, 13(4):559–576, 2005.

[3] A.M. Annaswamy and A.F. Ghoniem. Active control of combustion instability: Theory and practice. *IEEE Control Systems Magazine*, 22(6):37–54, 2002.

[4] K.J. Åström. Limitations on control system performance. *European Journal of Control*, 6(1):1–19, 2000.

[5] K.J. Åström and T. Hägglund. *Automatic Tuning of PID Controllers*. Instrument Society of America, Research Triangle Park, NC, 1988.

[6] K.J. Åström and T. Hägglund. *PID Controllers: Theory, Design, and Tuning*. Instrument Society of America, Research Triangle Park, NC, 2nd edition, 1995.

[7] K.J. Åström and T. Hägglund. The future of PID control. *Chem. Eng. Progress*, 9:1163–1175, 2001.

[8] K.J. Åström, C.C. Hang, and B.C. Lim. A new Smith predictor for controlling a process with an integrator and long dead-time. *IEEE Trans. Automat. Control*, 39:343–345, 1994.

[9] K.J. Åström and B. Wittenmark. *Computer-Controlled Systems: Theory and Design*. Prentice-Hall, Englewood Cliffs, NJ, 2nd edition, 1989.

[10] C.T.H. Baker, G.A. Bocharov, and F.A. Rihan. A report on the use of delay differential equations in numerical modelling in the biosciences. Numerical Analysis Report No. 343, Manchester Centre for Computational Mathematics, Manchester, England, August 1999. Available at http://www.maths.man.ac.uk/~nareports/narep343.ps.gz.

[11] J.A. Ball and C.M. Ran. Left versus right canonical Weiner-Hopf factorization. In I. Gohberg and M.A. Kaashoek, editors, *Constructive*

Methods of Wiener-Hopf Factorization, volume 21 of *Operator Theory: Advances and Applications*, pages 9–38. Birkhäuser Verlag, Basel, 1986.

[12] H. Bart, I. Gohberg, and M.A. Kaashoek. *Minimal Factorization of Matrix and Operator Functions*. Birkhäuser Verlag, Basel, 1979.

[13] T. Başar and P. Bernhard. H_∞-*Optimal Control and Related Minimax Design Problems: A Dynamic Game Approach*. Birkhäuser, Boston, 2nd edition, 1995.

[14] D.S. Bernstein. *Matrix Mathematics: Theory, Facts, and Formulas with Application to Linear Systems Theory*. Princeton University Press, NJ, USA, 2005.

[15] S. Bittanti, A.J. Laub, and J.C. Willems (Eds). *The Riccati Equation*. Springer-Verlag, Berlin, 1991.

[16] G.A. Bocharov and F.A. Rihan. Numerical modelling in biosciences using delay differential equations. *Journal of Computational and Applied Mathematics*, 125:183–199, 2000.

[17] J. Chen, L. Qiu, and O. Toker. Limitations on maximal tracking accuracy. *IEEE Trans. Automat. Control*, 45(2):326–331, 2000.

[18] P. Cominos and N. Munro. PID controllers: Recent tuning methods and design to specification. *IEE Proc. Control Theory Appl.*, 149(1):46–53, 2002.

[19] R. Curtain, G. Weiss, and M. Weiss. Coprime factorization for regular linear systems. *Automatica*, 32(11):1519–1531, 1996.

[20] R. Curtain, G. Weiss, and M. Weiss. Stabilization of irrational transfer functions by controllers with internal loop. In A.A. Borichev and N.K. Nikolskiĭ, editors, *Operator Theory: Advances and Applications*, volume 129, pages 179–207. Birkhäuser, 2001.

[21] R.F. Curtain and H. Zwart. *An Introduction to Infinite-Dimensional Linear Systems Theory*. Springer-Verlag, NY, 1995.

[22] P.H. Delsarte, Y. Genin, and Y. Kamp. The Nevanlinna-Pick problem for matrix-valued functions. *SIAM Journal on Applied Math.*, 36:47–61, 1979.

[23] C.N. Dorny. *A Vector Space Approach to Models and Optimization*. Robert E. Krieger Publishing Co. Inc., 1975. Chapters 1–5 are available at http://www.seas.upenn.edu/~dorny/VectorSp/vector_space.html.

[24] A.P. Dowling and A.S. Morgans. Feedback control of combustion oscillations. *Annu. Rev. Fluid. Mech.*, 37:151–182, 2005.

[25] J.C. Doyle, K. Glover, P.P. Khargonekar, and B.A. Francis. State-space solutions to standard H^2 and H^∞ control problems. *IEEE Trans. Automat. Control*, 34(8):831–847, 1989.

[26] H. Dym, T. Georgiou, and M.C. Smith. Explicit formulas for optimally robust controllers for delay systems. *IEEE Trans. Automat. Control*, 40(4):656–669, 1995.

[27] K. Engelborghs, M. Dambrine, and D. Roose. Limitations of a class of stabilization methods for delay systems. *IEEE Trans. Automat. Control*, 46(2):336–339, 2001.

[28] A. Fattouh, O. Sename, and J.-M. Dion. Pulse controller design for linear time-delay systems. In *The 1st IFAC Symposium on System Structure and Control*, Prague, Czech Republic, 2001.

[29] D.S. Flamm and S.K. Mitter. H^∞ sensitivity minimization for delay systems. *Syst. Control Lett.*, 9:17–24, 1987.

[30] C. Foias, H. Özbay, and A. Tannenbaum. *Robust Control of Infinite Dimensional Systems: Frequency Domain Methods*, volume 209 of *LNCIS*. Springer-Verlag, London, 1996.

[31] C. Foias, A. Tannenbaum, and G. Zames. Weighted sensitivity minimization for delay systems. *IEEE Trans. Automat. Control*, 31(6):763–766, 1986.

[32] B.A. Francis. *A Course in H_∞ Control Theory*, volume 88 of *LNCIS*. Springer-Verlag, NY, 1987.

[33] G.F. Franklin, J.D. Powell, and M.L. Workman. *Digital Control of Dynamic Systems*. Addison-Wesley, Reading, MA, 2nd edition, 1990.

[34] T.T. Georgiou and M.C. Smith. w-stability of feedback systems. *Syst. Control Lett.*, 13:271–277, 1989.

[35] T.T. Georgiou and M.C. Smith. Graphs, causality and stabilizability: Linear, shift-invariant systems on $L_2[0,\infty)$. *Mathematics of Control, Signals, and Systems*, 6:195–223, 1993.

[36] I. Gohberg, S. Goldberg, and M.A. Kaashoek. *Classes of Linear Operators*, volume II. Birkhäuser, Basel, 1993.

[37] G.C. Goodwin, S.F. Graebe, and M.E. Salgado. *Control System Design*. Prentice-Hall, Upper Saddle River, NJ, 2001.

[38] M. Green. H_∞ controller synthesis by J-lossless coprime factorization. *SIAM J. Control Optim.*, 30(3):522–547, 1992.

[39] M. Green, K. Glover, D. Limebeer, and J. Doyle. A J-spectral factorization approach to H^∞ control. *SIAM J. Control Optim.*, 28(6):1350–1371, 1990.

[40] M. Green and D.J.N. Limebeer. *Linear Robust Control*. Prentice-Hall, Englewood Cliffs, NJ, 1995.

[41] G. Gu, J. Chen, and O. Toker. Computation of $L_2[0,h]$ induced norms. In *Proc. of the 35th IEEE Conference on Decision & Control*, pages 4046–4051, Kobe, Japan, 1996.

[42] K. Gu, V.L. Kharitonov, and J. Chen. *Stability of Time-Delay Systems*. Birkhäuser, 2003.

[43] K. Gu and S.-I. Niculescu. Additional dynamics in transformed time-delay systems. *IEEE Trans. Automat. Control*, 45(3):572–575, 2000.

[44] K. Gu and S.-I. Niculescu. Further remarks on additional dynamics in various model transformations of linear delay systems. *IEEE Trans. Automat. Control*, 46(3):497–500, 2001.

[45] K. Gu and S.-I. Niculescu. Survey on recent results in the stability and control of time-delay systems. *Journal of Dyn. Sys. Meas. Con. -Trans. ASME*, 125:158–165, 2003.

[46] W.K. Ho, C.C. Hang, and L.S. Cao. Tuning of PID controllers based on gain and phase margin specifications. *Automatica*, 31(3):497–502, 1995.

[47] I.M. Horowitz. *Synthesis of Feedback Systems*. Academic Press, NY, 1969.

[48] G.E. Hutchinson. Circular casual systems in ecology. *Anal. New York Acad. Sci.*, 50:221–246, 1948.

[49] K. Ichikawa. Frequency-domain pole assignment and exact model-matching for delay systems. *Int. J. Control*, 41:1015–1024, 1985.

[50] IEEE. *ANSI/IEEE 754-1985, Standard for Binary Floating-Point Arithmetic*. IEEE, New York, NY, USA, August 1985.

[51] O. Iftime and H. Zwart. Nehari problems and equalizing vectors for infinite-dimensional systems. *Syst. Control Lett.*, 45(3):217–225, 2002.

[52] R. Izmailov. Analysis and optimization of feedback control algorithms for data transfers in high-speed networks. *SIAM J. Control Optim.*, 34(5):1767–1780, 1996.

[53] Y. Kannai and G. Weiss. Approximating signals by fast impulse sampling. *Mathematics of Control, Signals, and Systems*, 6:166–179, 1993.

[54] K. Kashima. *General solution to standard H^∞ control problems for infinite-dimensional systems*. PhD thesis, Kyoto University, Kyoto, Japan, 2005.

[55] K. Kashima, H. Özbay, and Y. Yamamoto. On the mixed sensitivity optimization problem for stable pseudorational plants. In *Proc. of the 4th IFAC Workshop on Time-Delay Systems (TDS'03)*, Rocquencourt, France, September 2003.

[56] K. Kashima and Y. Yamamoto. Equivalent characterization of invariant subspaces of H^2 and applications to optimal sensitivity problems. In *Proc. of the 42nd IEEE Conference on Decision and Control*, Hawaii, USA, December 2003.

[57] K. Kashima and Y. Yamamoto. General solution to standard H^∞ control problems for a class of infinite-dimensional systems. In *Proc. of the 44th IEEE Conf. on Decision & Control and European Control Conference ECC 2005*, pages 2457–2462, Seville, Spain, December 2005.

[58] H. Kimura. *Chain-Scattering Approach to H^∞ Control*. Birkhäuser, Boston, 1996.

[59] A. Kojima and S. Ishijima. Robust controller design for delay systems in the gap-metric. *IEEE Trans. Automat. Control*, 40(2):370–374, 1995.

[60] H. Kwakernaak. A descriptor algorithm for the spectral factorization of polynominal matrices. In *IFAC Symposium on Robust Control Design*, Prague, Czech Republic, 2000.

[61] P. Lancaster. *Theory of Matrices*. Academic Press, NY, 1969.

[62] P. Lancaster and L. Rodman. *Algebraic Riccati Equations*. Clarendon Press, Oxford, UK, 1995.

[63] A.J. Laub. A Schur method for solving algebraic Riccati equations. *IEEE Trans. Automat. Control*, 24(6):913–921, 1979.

[64] R.K. Lea, R. Allen, and S.L. Merry. A comparative study of control techniques for an underwater fight vehicle. *Int. J. of Systems Science*, 30(9):947–964, 1999.

[65] H. Logemann, R. Rebarber, and G. Weiss. Conditions for robustness and nonrobustness of the stability of feedback systems with respect to small delays in the feedback loop. *SIAM J. Control Optim.*, 34:572–600, 1996.

[66] T.A. Lypchuk, M.C. Smith, and A. Tannenbaum. Weighted sensitivity minimization: General plants in H^∞ and rational weights. *Linear Algebra and its Applications*, 109:71–90, 1988.

[67] M.C. Mackey and L. Glass. Oscilations and chaos in physiological control systems. *Science*, 197:287–289, 1977.

[68] R. Majumder, B. Chaudhuri, B.C. Pal, and Q.-C. Zhong. A unified-Smith-predictor approach for power system damping control design using remote signals. *IEEE Trans. Control Syst. Technology*, 13(6):1063–1068, 2005.

[69] A. Malinowski, T. Booth, S. Grady, and B. Huggins. Real time control of a robotic manipulator via unreliable Internet connection. In *The 27th Annual Conference of the IEEE Industrial Electronics Society, IECON'01*, volume 1, pages 170–175, 2001.

[70] A.Z. Manitius and A.W. Olbrot. Finite spectrum assignment problem for systems with delays. *IEEE Trans. Automat. Control*, 24(4):541–553, 1979.

[71] S. Mascolo. Congestion control in high-speed communication networks using the Smith principle. *Automatica*, 35(12):1921–1935, 1999.

[72] M.R. Matausek and A.D. Micic. On the modified Smith predictor for controlling a process with an integrator and long dead-time. *IEEE Trans. Automat. Control*, 44(8):1603–1606, 1999.

[73] A. Medvedev. Disturbance attenuation in finite-spectrum-assignment. *Automatica*, 33(6):1163–1168, 1997.

[74] G. Meinsma. *J*-spectral factorization and equalizing vectors. *Syst. Control Lett.*, 25:243–249, 1995.

[75] G. Meinsma and L. Mirkin. H^∞ control of systems with multiple I/O delays. In *Proc. of the 4th IFAC Workshop on Time-Delay Systems (TDS'03)*, Rocquencourt, France, September 2003.

[76] G. Meinsma and L. Mirkin. H^∞ control of systems with multiple I/O delays. Part II: Simplifications. In *Proc. of the 44th IEEE Conf. on Decision & Control and European Control Conference ECC 2005*, pages 5054–5059, Seville, Spain, December 2005.

[77] G. Meinsma and L. Mirkin. H^∞ control of systems with multiple I/O delays via decomposition to adobe problems. *IEEE Trans. Automat. Control*, 50(2):199–211, 2005.

[78] G. Meinsma, L. Mirkin, and Q.-C. Zhong. Control of systems with I/O delay via reduction to a one-block problem. *IEEE Trans. Automat. Control*, 47(11):1890–1895, 2002.

[79] G. Meinsma and H. Zwart. The standard H^∞ control problem for dead-time systems. In *Proc. of the MTNS'98 Symposium*, pages 317–320, Padova, Italy, 1998.

[80] G. Meinsma and H. Zwart. On H^∞ control for dead-time systems. *IEEE Trans. Automat. Control*, 45(2):272–285, 2000.

[81] W. Michiels, S. Mondié, and D. Roose. Necessary and sufficient conditions for a safe implementation of distributed delay control. In *Proc. of the CNRS-NSF Workshop: Advances in Time-Delay Systems*, pages 85–92, Paris, France, Jan. 2003.

[82] R.H. Middleton and G.C. Goodwin. Improved finite word-length characteristics in digital control using delta operators. *IEEE Trans. Automat. Control*, 31(11):1015–1021, 1986.

[83] R.H. Middleton and G.C. Goodwin. *Digital Control and Estimation: A Unified Approach*. Prentice-Hall, Inc., Englewood Cliffs, NJ, 1990.

[84] R.H. Middleton and S.F. Graebe. Slow stable open-loop poles: To cancel or not to cancel. *Automatica*, 35(5):877–886, 1999.

[85] L. Mirkin. On the extraction of dead-time controllers from delay-free parametrizations. In *Proc. of 2nd IFAC Workshop on Linear Time Delay Systems*, pages 157–162, Ancona, Italy, 2000.

[86] L. Mirkin. Are distributed-delay control laws intrinsically unapproximable? In *Proc. of the 4th IFAC Workshop on Time-Delay Systems (TDS'03)*, Rocquencourt, France, September 2003.

[87] L. Mirkin. On the extraction of dead-time controllers and estimators from delay-free parametrizations. *IEEE Trans. Automat. Control*, 48(4):543–553, 2003.

[88] L. Mirkin. On the H^∞ fixed-lag smoothing: How to exploit the information preview. *Automatica*, 39(8):1495–1504, 2003.

[89] L. Mirkin and N. Raskin. Every stabilizing dead-time controller has an observer-predictor-based structure. In *Proc. of the 7th IEEE Mediterranean Conference in Control and Automation*, pages 1835–1844, Haifa, Israel, 1999.

[90] L. Mirkin and N. Raskin. State-space parametrization of all stabilizing dead-time controllers. In *Proc. of the 38th Conf. on Decision & Control*, pages 221–226, Phoenix, Arizona, USA, Dec. 1999.

[91] L. Mirkin and N. Raskin. Every stabilizing dead-time controller has an observer-predictor-based structure. *Automatica*, 39(10):1747–1754, 2003.

[92] L. Mirkin and G. Tadmor. H^∞ control of system with I/O delay: A review of some problem-oriented methods. *IMA J. Math. Control & Information*, 19(1):185–199, 2002.

[93] L. Mirkin and Q.-C. Zhong. Coprime parametrization of 2DOF controller to obtain sub-ideal disturbance response for processes with dead time. In *Proc. of the 40th IEEE Conf. on Decision & Control*, pages 2253–2258, Orlando, USA, December 2001.

[94] L. Mirkin and Q.-C. Zhong. 2DOF controller parametrization for systems with a single I/O delay. *IEEE Trans. Automat. Control*, 48(11):1999–2004, 2003.

[95] A.A. Moelja and G. Meinsma. H_2-optimal control of systems with multiple I/O delays: Time domain approach. *Automatica*, 41(7):1229–1238, 2005.

[96] S. Mondié, R. Lozano, and J. Collado. Resetting process-model control for unstable systems with delay. In *Proc. of the 40th IEEE Conference on Decision & Control*, volume 3, pages 2247–2252, Orlando, Florida, USA, 2001.

[97] S. Mondié and W. Michiels. Finite spectrum assignment of unstable time-delay systems with a safe implementation. *IEEE Trans. Automat. Control*, 48(12):2207–2212, 2003.

[98] S. Mondié and O. Santos. Approximations of control laws with distributed delays: A necessary condition for stability. In *The 1st IFAC Symposium on System Structure and Control*, Prague, Czech Republic, August 2001.

[99] M. Morari and E. Zafiriou. *Robust Process Control*. Prentice-Hall, Inc., Englewood Cliffs, NJ, 1989.

[100] K.M. Nagpal and R. Ravi. H^∞ control and estimation problems with delayed measurements: State-space solutions. *SIAM J. Control Optim.*, 35(4):1217–1243, 1997.

[101] C.P. Neuman. Properties of the delta operator model of dynamic physical systems. *IEEE Trans. Systems, Man, and Cybernetics*, 23(1):296–301, 1993.

[102] S.-I. Niculescu. *Delay Effects on Stability: A Robust Control Approach*, volume 269 of *LNCIS*. Springer-Verlag, Heidelberg, Germany, 2001.

[103] S.-I. Niculescu, A.M. Annaswamy, J.P. Hathout, and A.F. Ghoniem. On time-delay induced instabilities in low order combustion systems. In *Proc. 1st IFAC Symp. Syst. Struct. Contr.*, Prague, Czech Republic, 2001.

[104] E. Nobuyama. Robust stabilization of time-delay systems via reduction to delay-free model matching problems. In *Proc. of the 31st IEEE Conference on Decision & Control*, volume 1, pages 357–358, Tucson, Arizona, 1992.

[105] E. Nobuyama. Robust deadbeat control of continuous-time systems. *Journal of the Society of Instrument and Control Engineers.*, 38(9):547–552, 1999.

[106] E. Nobuyama, S. Shin, and T. Kitamori. Deadbeat control of continuous-time systems: MIMO case. In *Proc. of the 35th IEEE Conference on Decision & Control*, pages 2110–2113, Kobe, Japan, 1996.

[107] J.E. Normey-Rico and E.F. Camacho. Robust tuning of dead-time compensators for process with an integrator and long dead-time. *IEEE Trans. Automat. Control*, 44(8):1597–1603, 1999.

[108] A.W. Olbrot. Stabilizability, detectability, and spectrum assignment for linear autonomous systems with general time delays. *IEEE Trans. Automat. Control*, 23(5):887–890, 1978.

[109] A.W. Olbrot. Finite spectrum property and predictors. *Annual Reviews in Control*, 24(1):125–134, 2000.

[110] J.C. Oostveen and R.F. Curtain. Riccati equations for strongly stabilizable bounded linear systems. *Automatica*, 34(8):953–967, 1998.

[111] H. Özbay, T. Kang, S. Kalyanaraman, and A. Iftar. Performance and robustness analysis of an H^∞ based flow controller. In *Proc. the 38th IEEE Conference on Decision and Control*, pages 2691–2696, Phoenix, Arizona, USA, 1999.

[112] Z.J. Palmor. Stability properties of Smith dead-time compensator controllers. *Int. J. Control*, 32(6):937–949, 1980.

[113] Z.J. Palmor. Time-delay compensation — Smith predictor and its modifications. In S. Levine, editor, *The Control Handbook*, pages 224–237. CRC Press, 1996.

[114] J.R. Partington. Some frequency-domain approaches to the model reduction of delay systems. *Annual Reviews in Control*, 28:65–73, 2004.

[115] J.R. Partington and K. Glover. Robust stabilization of delay systems by approximation of coprime factors. *Syst. Control Lett.*, 14:325–331, 1990.

[116] L. Qiu and E.J. Davison. Performance limitations of non-minimum phase systems in the servomechanism problem. *Automatica*, 29(2):337–349, 1993.

[117] P.-F. Quet, B. Atalar, A. Iftar, H. Özbay, S. Kalyanaraman, and T. Kang. Rate-based flow controllers for communication networks in the presence of uncertain time-varying multiple time-delays. *Automatica*, 38(6):917–928, 2002.

[118] P.-F. Quet, S. Ramakrishnan, H. Özbay, and S. Kalyanaraman. On the H^∞ controller design for congestion control in communication networks with a capacity predictor. In *Proc. the 40th IEEE Conference on Decision and Control*, pages 598–603, Orlando, Florida, USA, 2001.

[119] A.C.M. Ran. Necessary and sufficient conditions for existence of J-spectral factorization for para-Hermitian rational matrix functions. *Automatica*, 39(11):1935–1939, 2003.

[120] A.C.M. Ran and L. Rodman. On symmetric factorizations of rational matrix functions. *Linear and Multilinear Algebra*, 29:243–261, 1991.

[121] J.-P. Richard. Time-delay systems: An overview of some recent advances and open problems. *Automatica*, 39(10):1667–1694, 2003.

[122] R. Safaric, K. Jezernik, D.W. Calkin, and R.M. Parkin. Telerobot control via Internet. In *Proceedings of the IEEE International Symposium on Industrial Electronics*, volume 1, pages 298–303, 1999.

[123] O. Santos and S. Mondié. Control laws involving distributed time delays: Robustness of the implementation. In *Proc. of the 2000 American Control Conference*, volume 4, pages 2479–2480, 2000.

[124] F.G. Shinskey. *Process Control Systems*. McGraw-Hill Book Company, 1967.

[125] M.C. Smith. Singular values and vectors of a class of Hankel operators. *Syst. Control Lett.*, 12(4):301–308, 1989.

[126] O.J.M. Smith. Closer control of loops with dead time. *Chem. Eng. Progress*, 53(5):217–219, 1957.

[127] O.J.M. Smith. *Feedback Control Systems*. McGraw-Hill Book Company Inc., 1958.

[128] W. Su, L. Qiu, and J. Chen. Fundamental performance limitations in tracking sinusoidal signals. *IEEE Trans. Automat. Control*, 48:1371–1380, 2003.

[129] Z. Świder. Realization using the γ-operator. *Automatica*, 34(11):1455–1457, 1998.

[130] G. Tadmor. The Nehari problem in systems with distributed input delays is inherently finite dimensional. *Syst. Control Lett.*, 26(1):11–16, 1995.

[131] G. Tadmor. Robust control in the gap: A state space solution in the presence of a single input delay. *IEEE Trans. Automat. Control*, 42(9):1330–1335, 1997.

[132] G. Tadmor. Weighted sensitivity minimization in systems with a single input delay: A state space solution. *SIAM J. Control Optim.*, 35(5):1445–1469, 1997.

[133] G. Tadmor. Robust control of systems with a single input lag. In L. Dugard and E.I. Verriest, editors, *Stability and Control of Time-Delay Systems*, volume 228 of *LNCIS*, pages 259–282. Springer-Verlag, London, 1998.

[134] G. Tadmor. The standard H^∞ problem in systems with a single input delay. *IEEE Trans. Automat. Control*, 45(3):382–397, 2000.

[135] P.K.S. Tam and J.B. Moore. Stable realization of fixed-lag smoothing equations for continuous-time signals. *IEEE Trans. Automat. Control*, 19(1):84–87, 1974.

[136] K.K. Tan, Q.G. Wang, and C.C. Hang. *Advances in PID Control*. Springer, Berlin, 1999.

[137] E.G.F. Thomas. Vector-valued integration with applications to the operator-valued H^∞ space. *J. Mathematical Control and Information*, 14:109–136, 1997.

[138] O. Toker and H. Özbay. H^∞ optimal and suboptimal controllers for infinite dimensional SISO plants. *IEEE Trans. Automat. Control*, 40(4):751–755, 1995.

[139] O. Toker and H. Özbay. Gap metric problem for MIMO delay systems: Parametrization of all suboptimal controllers. *Automatica*, 31(7):931–940, 1995.

[140] M.S. Triantafyllou and M.A. Grosenbaugh. Robust control for underwater vehicle systems with time delays. *IEEE Journal of Oceanic Engineering*, 16(1):146–151, 1991.

[141] V. Van Assche, M. Dambrine, J.F. Lafay, and J.P. Richard. Some problems arising in the implementation of distributed-delay control laws. In *Proc. of the 38th IEEE Conference on Decision & Control*, pages 4668–4672, Phoenix, Arizona, USA, 1999.

[142] V. Van Assche, M. Dambrine, J.F. Lafay, and J.P. Richard. Implementation of a distributed control law for a class of systems with delay. In *Proc. of the 3rd IFAC Workshop on Time-Delay Systems*, pages 266–271, USA, 2001.

[143] P. Varaiya and J. Walrand. *High-performance Communication Networks*. Morgan Kaufmann Publishers, San Francisco, CA, 1996.

[144] Q.G. Wang, T.H. Lee, and K.K. Tan. *Finite Spectrum Assignment for Time-Delay Systems*. Springer-Verlag, London, 1999.

[145] Q.G. Wang, B. Zou, T.H. Lee, and Q. Bi. Auto-tuning of multivariable PID controllers from decentralized relay feedback. *Automatica*, 33(3):319–330, 1997.

[146] K. Watanabe. Finite spectrum assignment and observer for multivariable systems with commensurate delays. *IEEE Trans. Automat. Control*, 31(6):543–550, 1986.

[147] K. Watanabe and M. Ito. A process-model control for linear systems with delay. *IEEE Trans. Automat. Control*, 26(6):1261–1269, 1981.

[148] K. Watanabe, M. Ito, M. Kaneko, and T. Ouchi. Finite spectrum assignment problem for systems with delay in state variables. *IEEE Trans. Automat. Control*, 28(4):506–508, 1983.

[149] K. Watanabe, E. Nobuyama, T. Kitamori, and M. Ito. A new algorithm for finite spectrum assignment of single-input systems with time delay. *IEEE Trans. Automat. Control*, 37(9):1377–1383, 1992.

[150] K. Watanabe, E. Nobuyama, and A. Kojima. Recent advances in control of time delay systems—A tutorial review. In *Proc. of the 35th IEEE Conference on Decision & Control*, pages 2083–2089, Kobe, Japan, 1996.

[151] G. Weiss. Transfer functions of regular linear systems. Part I: Characterizations of regularity. *Trans. of the American Math. Society*, 342(2):827–854, 1994.

[152] G. Weiss, O.J. Staffans, and M. Tucsnak. Well-posed linear systems–a survey with emphasis on conservative systems. *Int. J. Appl. Math. Comput. Sci.*, 11(1):7–33, 2001.

[153] Y. Yamamoto. Learning control and related problems in infinite-dimensional systems. In H. Trentelman and J. Willems, editors, *Essays on Control: Perspectives in the Theory and its Applications*, pages 191–222. Birkhäuser, Boston, 1993.

[154] Y. Yamamoto, K. Hirata, and A. Tannenbaum. Some remarks on Hamiltonians and the infinite-dimensional one block H^∞ problem. *Syst. Control Lett.*, 29(2):111–117, 1996.

[155] C.-C. Yu. *Autotuning of PID Controllers: Relay Feedback Approach*. Springer Verlag, London UK, 1999.

[156] W. Zhang and Y.X. Sun. Modified Smith predictor for controlling integrator/time delay process. *Ind. Eng. Chem. Res.*, 35:2769–2772, 1996.

[157] Q.-C. Zhong. H^∞ control of dead-time systems based on a transformation. In *Proc. of the 3rd IFAC Workshop on Time Delay Systems*, Santa Fe, New Mexico, USA, December 2001.

[158] Q.-C. Zhong. Frequency domain solution to the delay-type Nehari problem using J-spectral factorization. In *Proc. of the 15th IFAC World Congress*, Barcelona, Spain, July 2002.

[159] Q.-C. Zhong. Control of integral processes with dead-time. Part 3: Deadbeat disturbance response. *IEEE Trans. Automat. Control*, 48(1):153–159, 2003.

[160] Q.-C. Zhong. Frequency domain solution to delay-type Nehari problem. *Automatica*, 39(3):499–508, 2003. See *Automatica* vol. 40, no. 7, 2004, p.1283 for minor corrections.

[161] Q.-C. Zhong. H^∞ control of dead-time systems based on a transformation. *Automatica*, 39(2):361–366, 2003.

[162] Q.-C. Zhong. On standard H^∞ control of processes with a single delay. *IEEE Trans. Automat. Control*, 48(6):1097–1103, 2003.

[163] Q.-C. Zhong. *Robust Control of Systems with Delays*. PhD thesis, Imperial College London, London, UK, Dec. 2003.

[164] Q.-C. Zhong. Robust stability analysis of simple systems controlled over communication networks. *Automatica*, 39(7):1309–1312, 2003.

[165] Q.-C. Zhong. Unified Smith predictor for dead-time systems. In *Proc. of the 4th IFAC Workshop on Time-Delay Systems (TDS'03)*, Rocquencourt, France, September 2003.

[166] Q.-C. Zhong. What's wrong with the implementation of distributed delay in a control law? In *Proc. of the 4th IFAC Workshop on Time-Delay Systems (TDS'03)*, Rocquencourt, France, September 2003.

[167] Q.-C. Zhong. On distributed delay in linear control laws. Part I: Discrete-delay implementations. *IEEE Trans. Automat. Control*, 49(11):2074–2080, 2004.

[168] Q.-C. Zhong. A case study for the delay-type Nehari problem. In *Proc. of the 44th IEEE Conf. on Decision & Control and European Control Conference ECC 2005*, pages 386–391, Seville, Spain, December 2005.

[169] Q.-C. Zhong. J-spectral factorization of regular para-Hermitian transfer matrices. *Automatica*, 41(7):1289–1293, 2005.

[170] Q.-C. Zhong. J-spectral factorization via similarity transformations. In *Proc. of the 16th IFAC World Congress*, Prague, Czech Republic, July 2005.

[171] Q.-C. Zhong. On distributed delay in linear control laws. Part II: Rational implementations inspired from the δ-operator. *IEEE Trans. Automat. Control*, 50(5):729–734, 2005.

[172] Q.-C. Zhong. On distributed delay in linear control laws. Part III: Rational implementation based-on extended bilinear transformations. Submitted for possible publication., 2005.

[173] Q.-C. Zhong. Rational implementation of distributed delay using extended bilinear transformations. In *Proc. of the 16th IFAC World Congress*, Prague, Czech Republic, July 2005.

[174] Q.-C. Zhong and H.X. Li. Two-degree-of-freedom PID-type controller incorporating the Smith principle for processes with dead-time. *Industrial & Engineering Chemistry Research*, 41(10):2448–2454, 2002.

[175] Q.-C. Zhong and L. Mirkin. Quantitative analyses of robust stability region and disturbance response of processes with an integrator and long dead-time. In *Proc. of the 40th IEEE Conf. on Decision & Control*, pages 1861–1866, Orlando, USA, December 2001.

[176] Q.-C. Zhong and L. Mirkin. Control of integral processes with dead-time. Part 2: Quantitative analysis. *IEE Proc. Control Theory Appl.*, 149(4):291–296, 2002.

[177] Q.-C. Zhong and J.E. Normey-Rico. Disturbance observer-based control for processes with an integrator and long dead-time. In *Proc. of the 40th IEEE Conf. on Decision & Control*, pages 2261–2266, Orlando, USA, December 2001.

[178] Q.-C. Zhong and J.E. Normey-Rico. Control of integral processes with dead-time. Part 1: Disturbance-observer based 2DOF control scheme. *IEE Proc. Control Theory Appl.*, 149(4):285–290, 2002.

[179] Q.-C. Zhong and G. Weiss. A unified Smith predictor based on the spectral decomposition of the plant. *Int. J. Control*, 77(15):1362–1371, 2004. Available at http://www.tandf.co.uk/journals/titles/00207179.asp.

[180] K. Zhou and J.C. Doyle. *Essentials of Robust Control*. Prentice-Hall, Upper Saddle River, NJ, 1998.

[181] K. Zhou, J.C. Doyle, and K. Glover. *Robust and Optimal Control*. Prentice-Hall, Englewood Cliffs, NJ, 1996.

[182] K. Zhou and P.P. Khargonekar. On the weighted sensitivity minimization problem for delay systems. *Syst. Control Lett.*, 8:307–312, 1987.

[183] H. Zwart, G. Weiss, and G. Meinsma. Prediction of a narrow-band signal from measurement data. In L. Caccetta, K.L. Teo, P.F. Siew, Y.H. Leung, L.S. Jennings, and V. Rehbock, editors, *Proc. of the 4th International Conference on Optimization: Techniques and Applications*, pages 329–336, Perth, Australia, 1998.

Index

2-degree-of-freedom   34, 139

algebraic Riccati equations   58
  block-diagram representation   60
  definition   58
  Hamiltonian matrix   58
  similarity transformations   61
    feedback   64
    inverse   65
    opposite   66
    parallel   62
    series   63
  solution generator   61
  stabilising solution   58
    grouping   68
    rank defect   67

bilinear transformation   207

cancellation property   198, 208
chain-scattering   46
  properties   49
  represenation   47, 121, 143
  state-space realizations   54
  transformations   47
co-prime factorisation   53, 140, 146
controller parameterisation   140, 164
convolution   176

δ-operator   196
delay   1
  control difficulties   22
  discrete   175
  distributed   173

examples of time-delay systems   2
  biosystems   8
  chemical processes   3
  combustion systems   6
  communication networks   3
  EGR systems   7
  shower   2
  underwater vehicles   4
  fundamental limitations   25, 28, 34
  transfer function   2
differential operator
  δ-operator   196
  γ-operator   208
distributed delay   *see* implementation of distributed delay

eigenspace
  antistable   77, 96
  stable   77, 96
entire function   175

finite-impulse-response (FIR) operators   45
  completion operator   45, 92
  truncation operator   45, 114, 116
finite-spectrum assignment   39
  observer–predictor representation   41
four-block problem   117
  delay-free   120
function
  δ 179
  pulse   176
  step   176

Index

function of matrix 181, 197
  power series 181
    convergency 181

γ-operator 207

H^2 problem 165
Hamiltonian property 58
hold filter 182

idempotent matrix 75
IEEE Standard 754 156
implementation of distributed delay 183
  rational implementation with δ-operator 197
    convergency 201
    example 204
    structure 203
  rational implementation with bilinear transformations 208
    convergency 211
    example 214
  with discrete delays
    convergency 187
    examples 188
    in the s-domain 185
    in the z-domain 183
integral processes with dead time 148
  subideal disturbance response 150
internal model control 24, 159

J-lossless 112
J-spectral co-factor 75
J-spectral factor 75
J-spectral factorisation 75, 77, 78, 92, 96
  numerical examples 84
J-unitary 112
Jordan form 157

$L_2[0, h]$-induced norm 72
limiting property 198, 208
linear fractional transformations 48
  homographic transformations (HMT)
    left $\mathcal{H}_l$ 49
    right $\mathcal{H}_r$ 49
    properties 49
    state-space realizations 56

  the standard LFT
    lower $\mathcal{F}_l$ 49
    upper $\mathcal{F}_u$ 49
    state-space realizations 55
LQ problem 167, 168

matrix
  idempotent 75
  nilpotent 75
  signature 77
  symplectic 62
Meinsma–Zwart idea 91
modified Smith predictor 30, 135, 142, 155
  Example 1 32
  Example 2 36
  implementation *see* implementation of distributed delay
  predictor–observer representation 40
  structure 31

Nehari Problem
  conventional 93
  delay-type 88
    numerical example 99
    optimal value 89
    performance range 89
    solution 89
    stable case 93
  extended delay-type 109, 123
    solution 110

one-block problem 119
  solution 124
output injection 52

para-Hermitian matrix 75
  equivalent condition 75
  general state-space form 76
  regular 76
periodic resetting mechanism 159
PID control 17
  Example 1 21
  Example 2 22
  structures 17
  tuning
    analytical tuning 20
    trial-and-error tuning 18
    Z-N tuning 19

predictor
    output predictor   43
    state predictor   43
projections   74, 77
    non-orthogonal   74
    orthogonal   75
    properties   75

quadrature rules   175
    rectangular   177, 182
    trapezoidal   177, 207

robustness analysis   145

sampled-data system   183
shift operator   208
Σ-matrix   68, 122
    definition   69
    properties   70
similarity transformation   57, 61, 77, 78, 80, 113, 167
simultaneous triangularisation   79
small-gain theorem   188
Smith predictor   24
    classical   24, 155
        disturbance response   27
        Example 1   28
        Example 2 (unstable)   30
        internal model control   24
        property   24
        robustness   25
        structure   24
    modified   *see* modified Smith predictor
    predicted state   39
    unified   *see* unified Smith predictor
stabilising controllers   39, 140, 164
    representations
        2DOF   142
        observer–predictor   41
        predictor–observer   40
stability
    dual-locus diagram   150
    exponential stability   165
    practical stability   26
    region   32, 36
    w-stability   26
stability margin

gain crossover frequency   20
gain margin   20
phase crossover frequency   20
phase margin   20
standard H^∞ problem   117
    delay-free, conventional   120, 133
    with a single I/O delay   119
        reducing to one-block problem   121
    solution   124
    transformed   130, 169
star product   54
static error   27, 31, 179, 197
static property   198, 208
symplectic matrix   62, 70
system
    cascading   49
    representations   46
        chain-scattering representation   *see* chain scattering
        input–output representation   47
    similarity transformation   *see* similarity transformation
    state-space operations   50, 94
        adjoint   51
        chain-scattering transformations   54
        co-prime factorisation   52, 53
        HMT   56
        inverse   51
        LFT   55
        output feedback   55, 56
        output injection   52
        parallel/addition   51
        piling up   51
        Redheffer star product   54
        series/cascade   51
        state feedback   52

Tustin's transformation   207

unified Smith predictor   156
    applications   164
    implementation   158
    structure   157

Zhou–Khargonekar formula   110
ZOH block   183, 185
    implementation in the s-domain   184